Choosing Whysly

Why we can't see what is right in front of us

Alex Bennet
Mountain Quest Institute

Cover graphic by Alex Bennet with AI DALL-E

FIRST EDITION (Softback)

MQIPress
Frost, West Virginia
303 Mountain Quest Lane, Marlinton, WV 24954
United States of America
Telephone: 304-799-7267
eMail: alex@mountainquestinstitute.com
www.mqresearchcenter.com
www.mountainquestinn.com

ISBN 978-1-968006-05-1

Dedication

This work is dedicated to the curious and courageous learners who dare to look beyond their personal blindness and bias to create the future imagined by our forefathers, rich in diversity, opportunity, and inclusion.

Deep Appreciation *to Robert Turner, Moria Levy, and Annie Green for their thoughts and energy that continue to spark and expand my thinking and writing. More recent appreciation of my AI partner, AI-PRO 4o, for support in scenario and story building and DALL-E for cover support.*

The etymology of *Whysly:* At Plato's Academy in the environs of ancient Rome, it was not sufficient to know *Why.* There, continuously inquiring with *Why* from myriad perspectives was the oracle reach for the unfolding of infinite wisdom. In some manner, *Whysly* seems whimsical in our new boundless exploration for intelligence. Emerging from Mountain Quest Institute, *Whysly* is inviting, profound in the amalgamation of terrestrial inquiry, and reaches for celestial wisdom to take us from the essence of the natural world to the precognition of human existence beyond all our past experience.

Preface

As the environment in which we live has increasingly imbibed fear and chaos, the results of fragmentation and hubris, I keep questioning how so many people can be oblivious to the tearing apart of our democratic way of life? Why do we choose not to see what is right in front of us?

This is a large question. Is it philosophical in nature? Certainly we can learn from neuroscience just how this works in the mind/brain. And surely there is research in psychology and sociology that addresses this. Or, perhaps theories around communication, culture or change studies? And how about critical theory? On the other hand, we certainly can learn in this regard from anthropology (the science of human beings), epistemology (the theory of knowledge), and education and pedagogy. Come to think of it, this idea of not seeing what is right in front of us may even bring in behavioral economics and political science, especially in today's volatile environment. And, frankly, how does spirituality fit into this why?

Reflecting, it is all of these fields and disciplines. It is only the human that puts things in boxes, applying labels (names) and then clustering like things together for ease of understanding. Only in a complex world, "like" things have many aspects to them, just as you and I do. We aren't one thing … and even when we are looking from a single perspective, other things are simultaneously going on both inside and outside of us with us learning and changing in response, whether we *know* it or not. We are complex adaptive systems that cannot live in stasis, which means we are either expanding or declining, living or dying, every moment of life.

So, what seems like a simple question becomes, well, complicated, no, complex, because it is about people and so it cannot help but be complex. In order to explore potential answers to this question of why we choose not to see what is right in front of us, we have to take a consilience approach, the "jumping together" of knowledge through the dynamic linking of facts and fact-based theories *across* fields to create a common groundwork of explanation.

When looking at the two cultures of science and humanism in 1959, C. P. Snow wrote about the importance of their interdependence. As he believed, "Nothing fundamental separates the course of human history from the course of physical history, whether in the stars or in organic diversity." (Wilson, 1998). In other words, the issues we face today demand that we use ALL of our resources—all of our understanding across a myriad of fields of learning—to produce a clearer view of reality.

As I moved through foundational references in the Mountain Quest 40,000+ volume research library exploring this theme, it became clear that this important topic was well-grounded. It was necessary to go back to the original research for working definitions and to discover the consistency that the expansion across various domains of knowledge was built upon. This was a fascinating learning journey. I've provided those references for anyone who chooses to dive deeper.

While exploring these foundational ideas and linking them with the current world turbulence, hubris, and fragmentation, the power and necessity of consilience to achieve understanding became even clearer. There is such diversity in human thought—just as there is magnificent diversity among people—all there to provide the opportunity for a higher level of learning and expansion. Why is it we can't see what is right in front of us? As a hypothesis, I would propose that, at some level, whether consciously or unconsciously, this has to do with choice. This hypothesis is explored throughout this book.

In deep diving recently deep into the synergy between critical thinking and creative thinking for another book, the realization emerged that during the last century, with a focus on strategic thinking, we had somehow been caught up in the how's that came along with technological advancement, to the detriment of focusing on the "why's." As we technologically continue to move forward, rapidly morphing toward a quantum mindset, past how's are left rapidly behind as new, exciting ideas and realities emerge right in front of us. For those of us who see them, who embrace them, it is critical to understand our why's in order to help us *whysly* choose—and create—our future. It is my hope this book will help us do so.

Alex Bennet

Contents

WHY?

Foreword

In her early years, my dear friend, Dr. Alex, drew deeply from her Greek family upbringing. Now, it is no vast stretch of the imagination that in *Whysly*, she transports us back two and a half millennia to the foundations of Greek wisdom. On this excursion, Alex shifts us to cognition cultivated in the outskirts of ancient Athens at Plato's Academy, where we find a genesis of the modern university. We ponder there, as it were, at the feet of Plato, the philosopher—teacher par excellence. For him, the meaning of existence, the purposes of life, the truth, and the beauty of what is before us and around us unfold delightfully as we pursue unending, bold, insightful questions—the *Why?* To be sure, with Plato, the questions are more important than the answers. As the questions shift and the responses accumulate, our perspectives shift and we see more; as the breadth and depth of our vision unfolds, it follows as the day the night, we also begin to know more *whysly*.

Plato and Dr. Alex want us to indulge our senses, our field of view, what it presents to us and how it lifts us. To that end, Alex basks in consilience—coalescing disparate perspectives together into new and revealing tapestries of knowledge. She wants us to see more and to know more—mixing fields and perceptions to illuminate what is dancing together, to envision it more clearly in our mind's eye. When we delight in the kaleidoscopic magnitude of these variations before us, we experience, as Plato offers in the dramatic dialogue *Theaetetus,* that "*Wonder is the beginning of wisdom.*"

Indeed, for Plato, at the Academy, there are never enough questions. Alex fills that void quite well. Do these endeavors reveal truth? That, somehow, is more dependent upon the seeker and the search than upon the aggregation of answers. What begins as steps or a series of questions can culminate in a vista offering a clearer perspective. However, it is not an easy ascent, and Alex offers an abundance of guideposts for the climb. You have endless choices.

Fast forward to Dr. Alex's world of the modern knowledge worker heralded by Fritz Machlup, the economist, and Peter Drucker, the management sage in the closing third of our last century, with 39 books in 36 languages on the new knowledge economy. By the end of the century, Alex and her partner and spouse, Dr. David Bennet, were in the thick of this new knowledge-centric world. David co-founded a firm that provided consulting services for managing complexity in this new knowledge era, and Alex served as the Chief Knowledge Officer and Deputy Chief Information Officer for Enterprise Integration in the U.S. Department of the Navy, one of the largest organizations in the world. Their combined efforts eventually created the Mountain Quest Institute (MQI) in the mountains of West Virginia, which has published over 40 books. All of this is to say that Plato's work at the Academy, which elucidated the nature of knowledge and learning, is being deeply replicated in the realm of multiple perspectives, formerly known as consilience.

A brief example to express the value of enhanced perspectives as championed at MQI. At the onset of the modern knowledge era, MacGregor Company management consultants popularized the Nine Dot Puzzle, which challenged people to connect nine dots in a format using three or fewer straight lines. As you find this puzzle on the Internet, the solution is achieved by "thinking outside the box". Once that creative window was introduced, the answers flowed, and the solutions exceeded 100. But who is counting?

Now, Dr. Alex is offering out-of-the-box perspectives that essentially make our potential unlimited. In this new era of intelligence, through MQI offerings* of Unleashing the Human Mind, Reblooming Knowledge, Whole Thought, Infinite Learning and Knowledge Capacities, and—specifically addressing current challenges—the exploration of fRAGmentation and Hubris leading up to this volume, the Nine Dot Puzzle has become three-dimensional. With the coalescing of Human Intelligence (HI) and AI, the solutions are infinite. The MQI *iFusion* endeavor offers that level of *Whysly* for us!

Robert Turner, Author, Researcher,
Mountain Quest Institute Associate

***MQI-published texts in support of iFusion:** (1) *Unleashing the Human Mind: A Consilience Approach to Managing Self* (2022); (2) *Reblooming the Knowledge Movement: The Democratization of Organizations* (2023); (3) *Innovative Creativity: Creating with Innovation in Mind* (2024); (4) *Whole Thought: The Rise of Human Intelligence* (2024); (5) *Knowledge Capacities I: Igniting Whole Thought* (2024); (6) *Infinite Learning: Your Life, Your Choice* (2025); (7) *Knowledge Capacities II: Cultivating Infinite Learning* (2024); (8) *fRAGmentation: The Power of e Pluribus Unum* (2025); (9) *The Taint of Hubris: Rediscovering Humility and Bridging Authentic Connection* (2025); and also in support (10) *C&C Thinking: Becoming Whole* (critical and creative) (2025); (11) *Contiguity: Entangled Living and Learning* (2025); and a five-book workbook series *University of the Universe* (Experiential, Existential, Environmental, Relational, and Temporal) (released throughout 2025).

Chapter 1
Introduction

In a world where chaos holds sway,
We seek out new truths every day.
With *Whysly* intent,
Our focus is bent
On horizons where insights will play.

In the complex landscape of human perception, the phenomenon of "not seeing what is right in front of you" forces us to confront the intricate interplay of cognitive processes, emotional biases, and social influences that shape our understanding of reality. In this usage, the verb "seeing" is both literal and metaphorical, representing sight as well as noticing, perceiving, comprehending, or knowing something beyond visual observation.

In Phenomenology, this "seeing" is a way of engaging with the world, an active process of meaning-making, rather than passive image reception. For example, in Plato's well-known allegory of the cave, "seeing the light" stands for intellectual enlightenment and recognition of truth, not just an exposure to visual images. Thus, when we explore "why we can't see what is right in front of us", we are exploring the "why" we are not fully and intellectually engaging with the world in which we currently exist.

Unfortunately, there's no simple answer to why this occurs. This is often hidden in the realms of our internal workings. For example, individuals often find themselves ignoring or denying evidence that contradicts their beliefs, largely due to deep-seated cognitive dissonance, cultural conditioning, and/or group dynamics.

This phenomenon transcends disciplinary boundaries, as insights from psychology reveal how cognitive biases and emotional tagging can skew perception, while sociological frameworks emphasize the role of socialization and ideology in shaping individual worldviews. Neuroscience adds another layer, illustrating how attention and emotional responses are governed by brain structures that prioritize certain information over others.

Moreover, educational theories highlight the importance of critical thinking and pedagogy in fostering greater awareness, while political science examines how ideologies and media framing can reinforce selective perceptions. The convergence of these disciplines, and others, deepens our understanding of why certain truths remain elusive to individuals—and sometimes entire communities—despite their apparent presence.

As we delve into this multifaceted issue, we will explore the underlying mechanisms that contribute to this state of selective blindness, drawing from a wealth of theoretical perspectives that illuminate the ways in which our minds and societies construct our perceived realities. By expanding our understanding of entangled insights, we aim to foster greater awareness of our individual blind spots and encourage conscious critical engagement with the truths currently shaping our lives and communities.

A World in Chaos: Economic and Social Turmoil in the U.S. (An Example)

In a nation where the principles of democracy and equality were once seen as hallmarks of society, the present moment finds many Americans in a state of uncertainty. Fractures of political divisiveness, deepening economic inequality, and widespread social unrest color the national mood. Recent years have seen a shift toward more extreme rhetoric, with ideological divisions sharpening and eroding a sense of shared reality. Policy debates increasingly hinge more on entrenched party lines and less on shared facts or nuanced reasoning.

Simultaneously, a series of high-impact decisions are made under the guise of fiscal necessity: essential social programs—Medicaid, Social Security, education subsidies, and public assistance—face deep cuts. Supporters of these measures point to budget shortfalls and rising national debt as reasons for "belt-tightening." Detractors, meanwhile, argue these are disproportionate sacrifices demanded mainly of the most vulnerable. The rationale of "returning tax benefits" is invoked, even as these tax breaks overwhelmingly favor the highest earners and largest corporations.

These moves raise profound ethical questions about the allocation of resources and the foundational values underlying policy. As critical services erode, those at the social margins—children, the elderly, low-income workers—face compounding hardships. Rising tariffs and inflation further restrict access to healthcare, education, and basic necessities, intensifying the gulf between privileged and precarious.

Proponents of austerity measures assert their necessity for economic "health," training efficiency for effectiveness. Critics document a steady deepening of inequality and loss of opportunity among disadvantaged populations. Language itself obscures impact: policy and media discussions frequently focus on "fiscal responsibility," glossing over human consequences and rarely centering on the lived experiences of those most affected.

Despite the justifications, long-term projections suggest these very cuts accelerate federal debt, ignoring the nuanced and interconnected effects, and contradicting stated fiscal objectives. Many citizens—overwhelmed by the chaos or worn down by relentless newsfeeds—remain unaware or actively avoidant of these consequences.

Misinformation, spin, and deepening partisanship fragment the information environment. Social media algorithms and echo chambers intensify this, pre-selecting the narratives users encounter and reinforcing ideological silos. Those who speak up against these trends encounter dismissal, sometimes hostility, as concerns are reframed as political attacks or dismissed as exaggeration. This selective blindness fosters business-as-usual, even as underlying social tensions intensify, threatening the underpinning of civic trust and cohesion, and democracy itself.

Communities now face spiraling healthcare expenses, rising poverty, and the fragmenting of public infrastructure and safety nets—challenges that demand collective awareness and coordinated response. The critical question emerges: How do ordinary people come to terms with and respond to a reality that is actively denied or minimized by prevailing narratives?

Though the magnitude of the challenge may seem overwhelming, it also brings an inflection point: the possibility for communities to re-examine their beliefs, engage critically, and reclaim the guiding ideas of democracy, dignity, and equity.

This example illustrates how current events can lead to chaos and disillusionment, especially when people fail to recognize the underlying issues due to framing, misinformation, or emotional attachment to dominant narratives. It invites reflection on how to foster greater awareness and understanding of these critical social issues.

Let's develop a fictional example that illustrates how a character can struggle to recognize the interconnectedness of their personal well-being with events occurring in their community, emphasizing the importance of awareness and empathy.

The Overlooked Horizon (A Story)

In the small town of Maplewood, Amanda Hughes sat at her kitchen table, the sunlight filtering through the window and casting warm circles on the wooden surface. Despite the inviting atmosphere, Amanda's heart was heavy with an unshakeable feeling of dissatisfaction. At 35, she had a stable job at a local bank, a modest home, and a loving family. However, a persistent longing gnawed at her: a sense of purpose and fulfillment that seemed to elude her.

Every morning, she rushed through her routine, putting on a brave face for her colleagues and family. Yet, beneath her cheerful demeanor, she felt a void. This feeling intensified as her community grappled with cuts in funding for essential services. Vital programs that once provided support for her neighbors—like food assistance and mental health services—had been slashed in recent years, all in the name of budget cuts and efficiency. Friends were losing their jobs. Yet, despite the visible impact of these changes, Amanda had convinced herself that they didn't affect her directly.

One evening, during a casual dinner with friends, the topic of budget cuts came up. "Have you heard they're cutting the community center's funding again?" her friend Sarah said, frowning. "It's heartbreaking. That place was a refuge for so many."

Amanda nodded, but a part of her felt detached. She had never been to the community center; it was just one of many buildings in town that didn't resonate with her. "I mean, I get it," she replied. "They need to save money, right? It's tough out there."

As her friends voiced their frustrations about the lack of support for the most vulnerable members of their community, Amanda couldn't relate. She had a comfortable life, after all. Her thoughts wandered to her own life—how busy she had been with work, family events, and social gatherings. Why waste energy worrying about things that didn't impact her directly?

That weekend, while grocery shopping, Amanda noticed the long lines at the local food pantry. Families with tired faces shuffled in and out, and the shelves looked nearly bare. She felt a pang of sympathy but quickly dismissed it. "They should just work harder," she thought. This sentiment seemed justified as her own family had worked hard to build their life.

But then, during a visit to the bank, her boss announced that the branch would be downsizing, and staff hours would be reduced. Amanda's stomach sank. Her pay would be cut, putting a dent in their finances. "I can manage," she told herself, but the unease bubbled beneath the surface.

Weeks passed, and as tensions within her community escalated, so did her silent struggle. Amanda began to feel the financial strain, yet she continued to soldier on, convincing herself that others had it worse. She organized family outings, participated in community events, and wore her cheerful facade like armor.

However, late at night, when the house was quiet, Amanda found herself scrolling through social media. Posts from friends echoed the frustrations she had ignored: stories of layoffs, personal struggles, and pleas for help. For a moment, she felt the weight of their experiences, but quickly shut down those nagging thoughts. "I'm fine," she whispered to herself. "I just need to work harder."

One fateful night, Amanda received a call from Sarah. The community was organizing a fundraiser to support the food pantry. "It's a great opportunity to help people, Amanda. We need as many hands as we can get," Sarah urged.

In a rare moment of clarity, Amanda hesitated. The invitation felt like a call to action, yet she wanted to find an excuse. "I really can't. I have a lot on my plate right now," she protested, even as her heart yearned to contribute.

That evening, Amanda sat alone in her living room, the weight of the decision hanging heavily above her. Her gaze fell on the photos lining the wall—moments of joy shared with family and friends. Yet, something felt incomplete, as if she was missing a connection to something deeper.

In that quiet moment, realization washed over her like a tide. She had been so focused on her own life, her own struggles, that she had ignored the plight of others and the impact that the larger social issues had on her community—including herself. While she had felt somewhat secure, the challenges faced by her neighbors were all interconnected.

The next day, with newfound determination, Amanda called Sarah. "I changed my mind. I want to help with the fundraiser," she said, her voice steady. This time, she would not only show up for herself but for her community.

As she joined her friends, a sense of belonging enveloped her. Amanda began to understand that the emotional fulfillment she had been craving was not found in isolation or attachment to her own circumstances. It was in the shared struggles, the connections, and the collective efforts to lift one another that she would find her true sense of purpose. Connected events were becoming clearer. She and her community were in this together.

In overcoming her own reluctance to see the broader picture, Amanda took the first step toward reconnecting with her community—and herself. Through her journey, she discovered that sometimes the most profound revelations come from addressing the needs of others, ultimately leading to personal fulfillment and a deeper understanding of the world surrounding her.

Amanda's journey toward seeing the larger reality began with choice—her choice—to get involved. Our journey toward understanding why we don't see what is right in front of us also begins with choice, to gain an understanding of exactly what that entails.

Chapter 2
Choice and Agency

In the realm where choices arise,
Our beliefs shape what's wise in our eyes.
Beyond what we see,
Lies agency free
As perceptions give way to surprise.

When considering human agency, it is tempting to imagine ourselves as independent choosers—rational agents who freely survey options and select among them. Yet in practice, our acts of choosing are profoundly shaped, perhaps even constrained, by our own wants, perceptions, and beliefs. Often, we are "blind" to realities outside our inner frameworks. In this sense, we exercise agency not as free and open arbiters of possibility, but as beings whose horizons of choice are circumscribed, sometimes invisibly, by the limits of our subjectivity.

Do we truly choose, or do we simply select among what we *want*—and perhaps even, what we *wish*—to see? Put another way: To what extent is our agency an act of "choosing to see what we choose to see," and what does this mean for the nature and scope of human freedom?

The Role of Beliefs in Human Choice

Human agency—the ability to make choices and act upon them—is deeply influenced by the beliefs an individual holds. Far from being free-floating acts of will, our choices are powerfully shaped, constrained, and sometimes even determined by the frameworks of perception and understanding we bring to the world.

Humans often imagine knowledge as a set of *objective* facts about the world. However, a growing body of philosophical, psychological, and neuroscientific research suggests that *our knowledge is unavoidably shaped by subjective filters*, focusing our attention on certain options while rendering others invisible.

A primary filter is our belief system. For example, Festinger's theory of cognitive dissonance (1957) shows that people are motivated to maintain consistency between their actions and beliefs, often choosing paths that minimize psychological conflict, even if it means ignoring alternative information. Similarly, Kunda (1990) describes "motivated reasoning" as our tendency to interpret facts in ways that protect our existing beliefs and desires, which in turn channels our decision-making along well-worn tracks.

From a more philosophical and constructivist perspective, Kelly (1955) argues that individuals deploy "personal constructs"—essentially, belief systems—to interpret reality and navigate choices. In this sense, our beliefs not only filter our perceptions but help build the very world in which our choices make sense.

Kuhn (1962) argues that even scientific knowledge is shaped by paradigms—community-wide belief systems that determine what scientists see and how they interpret their findings. Neuroscience brings further depth to this picture. According to Friston's (2010) "free-energy principle," the brain is a prediction machine: what we perceive and act upon is not simply given by the senses, but is shaped by prior beliefs and expectations. Our brains generate models of the world rooted in belief, and we choose largely within the boundaries those models afford.

Philosophical inquiry further complicates the story. Bernard Williams (1973a, 1973b) asks whether beliefs themselves are subject to choice, concluding that while agency is central to being human, consciously "deciding to believe" is limited. This is because beliefs are shaped by what we recognize as reasons or evidence, which themselves are subjective, filtered through our existing frameworks.

Lastly, ubiquitous cognitive biases, such as *confirmation bias*, which is detailed later in this book, reinforce the tendency to attend to and seek information that matches what we already believe (Nickerson, 1998). In practice, this means that rather than surveying all possible options, humans interpret new information within our existing frameworks, gravitating toward those options consistent with prior convictions, effectively narrowing the field of real choices. New facts are often accepted or rejected based on whether they fit with what we already believe.

A related second filter is *perception*. What we know begins with what we perceive, and perception itself is selective—shaped by expectations, emotions, attention, and prior experience. This brings in a third filter, *emotional influences*. Emotions direct what we find important, memorable, or meaningful. They color not only how we perceive events but how we remember and judge them. A fourth filter is *social and cultural framing*. What counts as knowledge—and how it's valued—is affected by cultural norms, language, and social context.

There is a depth of research that supports the connection between the subjectivity of knowledge, the roles of emotion, belief, and cultural context, and how they all relate to agency. The table below provides a brief summary of major themes running through this research.

Theme	Key Reference	Main Point
Scientific paradigms	Kuhn (1962)	Scientific knowledge is paradigm-bound, not strictly objective.
Personal constructs	Kelly (1955)	We interpret reality through idiosyncratic belief systems.
Emotion and cognition	Damasio (1994)	Emotion is central to forming knowledge and making choices.
Motivated reasoning, bias	Kunda (1990; Nickerson (1998)	Knowledge is shaped by desire, belief, and selective attention.
Subjective perception	Merleau-Ponty (1945/1962)	All perception and knowledge is embodied and subjective.
Cultural/social framing	Berger & Luckman (1966)	What "counts" as knowledge is determined socially and culturally.
Agency & reflective freedom	Taylor (1985)	Agency is meaningful through self-reflection on one's frameworks.
Dynamic nature of knowledge construction	Merleau-Ponty (1945/1962); Bennet, Bennet, & Turner (2022)	Knowledge is subjectively re-created in the instant for the situation and context at hand.
Interplay between cognition and emotion	Damasio (1994); Greenfield (2000)	All information coming in through the senses is subject to a subjective emotional tag.

Connections to Agency

If all knowledge is subjective, then so is the field of possible choice. What we believe we can do, what outcomes we see as likely or worthwhile, and even what options we recognize as available, are all constructed through our ever-changing, subjective frameworks. Agency is not a matter of unfiltered will acting upon pure reality—it is the capacity to choose *within the world as we know and feel it.*

The implications are multiple. First, the *boundaries of agency*. Agency is only as broad as the knowledge we possess and the ways we are able to frame and interpret the world. If we are blind to certain facts, perspectives, or possibilities, those choices do not exist in our field of agency.

Second, the *potential for growth*. While all knowledge is subjective, *being aware* of this subjectivity gives us the chance to expand our horizons—to seek out new experiences, perspectives, and emotions that alter our framework, thus broadening our possible choices.

Third, *responsibility and humility*. Recognizing the subjectivity of knowledge should lead to increased humility in asserting our choices and greater openness to others' perspectives.

This begs the question: *Is agency meaningful if it is so fundamentally subjective?* Let's explore both sides (the "yes" and "no") of that question, and perhaps look at a middle ground as well.

The Case for "Yes": Agency Remains Meaningful. Even if agency is radically subjective—filtered entirely through one's own beliefs and perceptions—it can still be deeply meaningful, both individually and socially.

- *Personal Significance:* From the inside, agency is about the felt experience of making choices—deliberating, intending, acting. This experience is foundational to how humans understand themselves as persons ("I am the one who decides…"), and how they assign responsibility and meaning to their lives.
- *Social and Moral Responsibility:* Societies are structured around the assumption that people are responsible for their actions, even if shaped by beliefs. Legal and ethical systems rely on the premise that, within the world as each person sees it, they are making choices.
- *Value of Subjective Worlds:* Meaning is not found only in some "objective" sense but also in the worlds we each inhabit. If my choices have meaning within my subjective worldview, that meaning is real for me, and for others who recognize it.

The Case for "No": Agency is Problematic if Radically Subjective. If agency is only meaningful within the limits of subjective belief, there are concerns.

- *Limits of Freedom:* If I cannot see beyond my beliefs, are my actions truly "free" or simply the outworking of conditioning and bias? Radical subjectivity could mean human agency is an illusion—people are pushed by inner forces they neither chose nor see.
- *Diminished Moral Accountability:* If individuals are entirely bound by their subjective frames, holding them responsible may seem unfair—could they really have chosen otherwise?
- *Social Fragmentation:* Different subjective frameworks can lead to conflicting truths. If agency is only meaningful relative to personal beliefs, the basis for common, shared understanding weakens.

Middle Ground: Recognizing Subjectivity but Seeking Expansion. Many philosophers and psychologists suggest that while agency is inevitably subjective, meaningful agency emerges when individuals are able to *reflect on*, *question*, and sometimes *expand* or *challenge* their beliefs.

- The process of *reflection*, *learning*, and *growing awareness* can allow people to modify their subjective frameworks—deepening the authenticity and range of their agency.
- Agency, then, is meaningful to the extent that individuals can recognize their filters and strive to transcend, revise, or at least acknowledge them.

Here's a summary table:

Position	Is Agency Meaningful?	Key Arguments
Radical Subjectivity	Yes	Agency is meaningful within subjective experience; personal worlds matter.
	No	If choices are predetermined by beliefs, "free" agency is an illusion.
Reflective Subjectivity	Yes, with qualification	Agency gains meaning as one becomes more aware of, and can revise, one's own beliefs.

While agency is meaningful, its meaning appears to change depending on *whether we see subjectivity as a prison or as a field for growth*. As long as individuals can reflect on and potentially broaden their perspectives, agency retains significant, even if its exercise will always be shaped by the limits of individual perception and belief.

In summary, human choice operates within a web of beliefs that shape not just what is chosen, but what is even *seen as* choosable. Our agency is thus both empowered and limited by belief: we are free to act, but we look (and choose) through the lenses our experiences and convictions have fashioned.

Can We Choose Beyond Our Limited Perceptions/Beliefs?

Let's frame our question clearly. If all agency (the ability to choose) operates within what we can perceive and believe, is it ever possible for a person to select something that lies *outside* those boundaries? Can we, by some act of will, transcend the invisible walls built by our current knowledge and worldview?

1. **The Sense in Which We Cannot.** From a phenomenological or psychological standpoint, *the limits of perception and belief are, by definition, the limits of choice.* Add to our conversation above, we cannot

consciously choose something that does not even appear as an option, or that we cannot conceptualize as possible, real, or desirable. Merleau-Ponty (1962/1945) argues that our lived, embodied experience structures all meaning and possibility. If something isn't within our horizon of sense, we cannot select it. Rumsfeld's (2002) famous "unknown unknowns" line is "What we don't know we don't know is unavailable to our deliberations" encapsulates the idea that there are factors affecting our decisions and understanding of which we are completely unaware. For example, a person raised in a culture that has no concept of romantic love cannot choose a love marriage. As a second example, someone who truly believes a wall is impassable cannot choose to walk through it.

2. **The Sense in Which We Can**. *Change is possible*—people's belief systems and perceptions do sometimes shift, allowing "new" choices to appear. But this process is often gradual and "outside-in," rather than a direct act of will. Mechanisms that expand possibility include:

- *Encountering the Unexpected:* New experiences can disrupt current beliefs and cause paradigm shifts (Kuhn, 1962). For example, a person may meet someone with a radically different worldview or witness a counterevidence event, causing them to question their old limits.
- *Education and Dialogue:* Exposure to new ideas through travel, learning or conversation can bring unseen options into view (Gadamer, 1960/1975).
- *Reflective Self-Doubt*: Crisis, doubt, or inner conflict, that which creates cognitive dissonance (Festinger, 1957) may prompt questioning and allow a person to entertain alternatives previously ignored.
- *Imagination and Art:* Creative expression sometimes helps people "try on" new ways of seeing, briefly transcending habitual belief.
- *BUT* ... in each case, it is *new perception/belief* that makes new choices possible—not direct selection beyond the limits of the old. We do not reach completely outside our worldviews by fiat, but by disruption, expansion, or transformation of them.

3. **The Role of Agency in Self-Transcendence**. Some philosophers (Taylor, 1985) suggest agency consists not just in "choosing among" options, but in the capacity to *reflect on* and *revise* the frameworks that structure choice in the first place. This means agency is exercised at two levels:

- Choosing within a framework, or

- Seeking to expand or reconsider the framework itself.
- BUT … even such expansion requires some perception that one's beliefs might be partial, limiting, or open to question.

In conclusion, generally a person cannot "choose" in the fullest sense *beyond* their current limits of perception and belief—those limits define the field of choice. But agency is meaningful partly *because* frameworks are not perfectly fixed: new experiences, encounters, and reflections can, sometimes unpredictably, expand what we can see, believe, and therefore choose. Agency includes not only selection within a worldview, but—crucially—the (always partial) possibility of changing that worldview itself.

Does Recognizing Our Subjective Limits Alter Our Agency?

In Plato's "Apology" 36c-38a, Socrates expresses that "the unexamined life is not worth living," emphasizing the importance of self-reflection and seeking wisdom such that one's beliefs and actions gain true value and purpose. The examined life, while more complex, also has greater scope for growth, meaning, and authentic agency. Let's explore both sides of this question along with its transformative potential.

1. **Recognizing Subjectivity as Agency-Enhancing**. *Awareness breeds possibility*. When we realize our knowledge and choices are filtered by subjective beliefs, emotions, and perspectives, we gain an opportunity for *meta-cognition*—thinking about our thinking. This can:

   - *Expand our options:* Recognizing the boundaries of our worldview makes it possible to seek out new perspectives, challenge our own biases, and learn from others. The *veil* that limits our choices becomes, paradoxically, more transparent.
   - *Foster humility and openness:* When we understand our limitations, we become more receptive to alternative viewpoints, feedback, and even radically different experiences.
   - *Increase reflective freedom:* As philosopher Charles Taylor (1985) argues, agency is not just about acting but also about reflecting on the frameworks within which we act. Recognizing subjectivity allows us to question and sometimes revise those very frameworks, thereby making new forms of agency possible.
   - *Example:* A scientist aware of how paradigms shape her research can intentionally seek cross-disciplinary insights, opening up creative and

innovative solutions unavailable within her original paradigm (Kuhn, 1962).

2. **Recognizing Subjectivity as Agency-Limiting.** Awareness can also breed paralysis or skepticism.

    - *Analysis paralysis:* Becoming highly aware of the ways our choices are limited or constructed can make decision-making harder, sometimes leading to doubt or hesitancy.
    - *Epistemic humility taken to extremes:* If we become so skeptical of our own perspective (radical subjectivism or relativism), we may feel that truly "authentic" or "informed" agency is impossible.
    - *Example:* A person paralyzed by self-doubt about their judgment may struggle to take action, worried any choice is simply the outcome of invisible bias.

3. **Transformative Potential**: *Changing the Nature of Agency.* Rather than simply increasing or decreasing agency, *recognizing subjectivity transforms its nature.*

    - *From naïve to reflective agency:* Instead of acting within our worldview by default, we become capable of shaping, questioning, and sometimes transcending elements of it. Dewey (1916/1933) argued that growth in self-knowledge is central to education and moral agency.
    - *Recursive self-improvement*: Metacognition—the ability to reflect on our own cognition—lets us improve not only our choices, but our *ways of choosing* (Flavell, 1979).

4. **Synthesis.** Recognizing the subjectivity of our perception and belief:

    - *Expands agency* by opening new options, fostering growth, and increasing the adaptability of individuals.
    - *May momentarily limit agency* by opening new options, fostering growth, and increasing the adaptability of individuals.
    - *Transforms agency* by opening new options, fostering growth, and increasing the adaptability of individuals.

After Thoughts

In essence, human agency is inextricably linked to the beliefs, perceptions, emotions, and cultural frameworks that shape our experience of the world. Far from being purely free, our choices emerge within horizons set by what we perceive and believe—limits that are both real and, at times, invisible. Yet, awareness of this subjectivity can itself become a source of expanded freedom, prompting self-reflection, humility, and the ongoing potential to reshape our frameworks. Ultimately, while we cannot always see *beyond* our present limits, agency finds its greatest significance in our capacity to recognize, question, and sometimes *transcend* them.

True freedom is not the absence of limits, but the awareness of them. By examining the lenses through which we "see", we open ourselves to growth, and, in doing so, widen the field of real choice. In the end, agency lies not just in choosing, but in the courage to question the world in which our choices arise.

Chapter 3
Breaking Through

To break through the boundaries we know,
We *Whysly* let old notions go.
With reappraisal in mind,
New dialogues find
Transformations that help us to grow.

Up to this point, we have seen how the boundaries of our beliefs and perceptions shape the horizons of our agency and the range of choices available to us. But what happens when we try to move beyond those limits? This section invites us to reflect on what it means to expand—or even break—the boundaries of our subjective worldview. We'll explore how individuals can gradually widen their horizons, experience disruptive shifts in their understanding, and navigate the challenges and rewards of profound belief change. Through practical processes and real-world examples, we'll examine both the opportunities and emotional complexities involved in transforming the very frames that once defined our sense of possibility.

What Would It Mean to Expand (or Break) These Boundaries?

Together, let's join in a reflective exploration of what it might mean to expand or break the boundaries of one's subjective perception and belief—and how that might impact agency and life as a whole.

1. **Expansion:** *Gradually Widening Our Horizon*. To *expand* our boundaries means to *grow the range* of what we can perceive, believe, and thus choose. This process is often incremental, involving:

   - *Exposure to New Experiences:* Traveling, meeting people from different backgrounds, or encountering unfamiliar ideas—these experiences introduce new perspectives and possibilities into our worldview.

   - *Engagement with Art and Literature:* Imaginative works enable us to vicariously experience worlds and mindsets other than our own, stretching our sense of what is possible and meaningful.

   - *Critical Reflection and Self-*Questioning: Through introspection, philosophical inquiry, or dialogue, we notice the limitations of our "natural attitude" and intentionally seek to revise or broaden it (*Socratic self-examination*, *Deweyan reflective thinking*). Socratic self-examination is the practice of critically questioning one's own beliefs and assumptions to seek clarity and truth, inspired by Socrates' method of inquiry. (See Appendix A.) As a second example, Deweyan reflective thinking is the process of thoughtfully considering a problem or experience, evaluating evidence, and reasoning systematically to reach a conclusion, as emphasized by John Dewey (1933).

- *Infinite Learning and Education:* Deep learning is not just accumulation of facts, but transformation: recognizing and challenging assumptions, entertaining ambiguous or paradoxical views, and integrating new knowledge (Bennet, 2025).
- *Result:* As these activities widen our perceptual/belief "window," more choices and forms of agency appear. Life becomes richer, more complex, sometimes more uncertain, but also there is more freedom.

2. **Breaking:** *Radical Disruption or Transformation*. To *break* the boundaries goes beyond expansion—*it is a fundamental overhaul, a rupture* in the very frame of understanding.
    - *"Conversion" or "Paradigm Shift":* As described by Kuhn (1962), sometimes a person or community abandons an old worldview in favor of a radically incompatible one (e.g., Copernican revolution, religious conversion, a personal "epiphany").
    - *Epistemic Crisis:* A sudden shock (trauma, contradiction, or revelation) can shatter long-held beliefs, forcing rapid reorganization of perception ("the scales fall from one's eyes"). William James (1902) speaks of "second birth" moments.
    - *Transcendence:* Mystical or "peak" experiences (Maslow, 1964) occasionally grant radically new perspectives—sometimes permanent, often fleeting.
    - *Result:* Breaking boundaries can feel like "waking up" into a new reality. This may open up new forms of agency, creativity, and empathy—but can also bring disorientation, identity crisis, or uncertainty.

3. **Implications:** *Ask: What are the implications?*
   - *Ethical and Social Growth:* As individuals expand or break boundaries, they become more capable of empathy, understanding, and moral responsibility—less bound by prejudice or closed-mindedness (Berger & Luckmann, 1966).
   - *Personal Creativity and Flexibility:* A wider or flexible worldview is strongly correlated with resilience, creativity, and the ability to thrive in complex or changing environments.
   - *Greater, but Hard-Won, Freedom*: Authentic agency becomes possible—to shape not just actions, but the frames through which actions and choices are understood.
4. **Limits and Cautions**:
   - *Never Fully "Transparent":* No matter how much our perspectives expand, we always remain within *some* horizon; we can never see with God's-eye objectivity (Gadamer, 1975/1960).
   - *Pain and Disorientation:* Breaking boundaries can be existentially unsettling—leading to periods of confusion, loss of identity, or emotional difficulty.

To expand or break the boundaries of one's perception and belief is to become capable of new choices, relationships, and authentic forms of selfhood. It is both the origin of growth and creativity, and a source of risk and disorientation. In becoming aware of and moving beyond old frames, we not only change *what* we can choose but *who* we choose to become.

Breaking Through These Boundaries

Here are three simple step-by-step processes to help break through the limits of one's perception and beliefs. Each is rooted in a distinct theoretical perspective.

Process 1: Harnessing Cognitive Reappraisal

Perspective: Neuroscience and cognitive psychology show that our beliefs, perceptions, and emotional reactions are not fixed, but can be consciously reconstructed through deliberate mental effort—a process called "cognitive reappraisal" or "reframing" (Ochsner & Gross, 2005; Siegel, 2010).

Context for Scenario Example: Sophie, an experienced elementary school teacher, is assigned a new principal who introduces unfamiliar teaching methods and provides unexpected critical feedback.

Steps

Step 1: Recognize the Trigger. Notice when a belief, perception, or reaction is automatic—especially in moments of strong emotion or habitual judgment.

Scenario: Sophie receives an email from the principal questioning her classroom management approach. She feels instantly angry and thinks, "This new principal has no idea what she's talking about."

Step 2: Pause and Observe. Use mindfulness techniques to stop and observe the underlying thought or interpretation, without immediately accepting it as true.

Scenario: Remembering her mindfulness training, Sophie pauses. She mentally notes, "I'm feeling angry. I'm jumping to conclusions."

Mindfulness Techniques. Two simple mindfulness techniques that can help you step back from automatic thoughts and observe them more objectively are focused breathing and a body scan.

Focused Breathing: Take a moment to focus on your breath. Find a comfortable seated position, close your eyes if you feel comfortable, and take a deep breath in through your nose, letting your abdomen expand. Hold the breath for a moment, then slowly exhale through your mouth. Repeat this process for about 3-5 breaths, allowing your mind to settle and observing any thoughts that arise without judgment. Notice how your body feels and any emotions present, holding space for them without trying to change them.

Body Scan: This technique involves bringing awareness to different parts of your body. Start at the top of your head and slowly move down to your toes. As you focus on each part, notice any tension or sensations. If you encounter a negative thought or emotion, acknowledge it and then visualize releasing it with each exhale. This practice helps ground you in the present moment and can create clarity around your feelings and reactions.

Step 3: Ask for Evidence. Deliberately question: "Is this belief or perception accurate? Is there evidence for an alternative view?" NOTE: Your AI partner can come in useful in Steps 3 and 4.

Scenario: Sophie asks herself, "Is it possible this feedback has value, or am I just threatened by change? Have I noticed these issues myself?"

Step 4: Reframe the Situation. Deliberately question: "Is this belief or perception accurate? Is there evidence for an alternative view?"

Scenario: Sophie considers, "Maybe this is an opportunity to learn, not a personal attack. What if I pilot these suggestions to see the impact?"

Step 5: Act on the New Frame. Deliberately question: "Is this belief or perception accurate? Is there evidence for an alternative view?"

Scenario: Sophie tries one of the new teaching strategies the following week, tracking student engagement and behavior.

Step 6: Reflect and Reinforce. Afterward, reflect on the outcome—has your perception shifted? Reinforce the new learning through journaling or sharing with others.

Scenario: Reviewing her notes, Sophie finds participation and focus slightly improved. She journals, "Trying something new wasn't easy, but I learned more about my students." She feels her professional identity expand.

A Real-World Example: You receive critical feedback at work and immediately feel defensive, thinking: "I'm failing."

Application:

- *Notice Trigger.* Recognize heightened emotion (defensiveness).
- *Pause & Observe.* Take a breath, mentally note your self-talk.
- *Ask for Evidence:* Ask, "Is it true I'm failing, or is this one area for growth?"
- *Reframe:* Consider, "Feedback is a chance to improve, not a sign I'm a failure."
- *Act on New Frame:* Ask clarifying questions rather than reacting defensively.
- *Reflect & Reinforce:* Note the improvement in your response; journal what you learned.

Here are a few tips: (1) Practice mindfulness—notice emotions without immediately acting on them. (2) Label thoughts as thoughts ("I am having the thought that …" rather than "I am …"). (3) Gather facts—what *other* explanations are possible? (4) Try a different response—if you'd usually withdraw, try engaging. (5) Keep a daily journal to reinforce new perspectives and spot patterns.

Process 2: Engaging in Dialogical Interpretation

Perspective: Hermeneutics stresses that shifting one's beliefs/perceptions often happens through real engagement with "the other"—in deep conversation or dialogue—where pre-existing assumptions are challenged and horizons are expanded (Gadamer, 1975/1960; Bohm, 1996).

Context for Scenario Example: Javier, a staunch environmentalist in his twenties, believes all farmers resist climate regulations. He meets **Debra**, a fifth-generation rancher, at a community town hall.

Steps

Step 1: Acknowledge Your Horizon. Admit you see from a particular cultural, historical, political, or psychological standpoint.

Scenario: Javier reminds himself, "I've mostly met city-dwellers—I barely know anyone who works the land."

Step 2: Seek Out Dialogue. Intentionally converse with someone holding a perspective, background, or expertise significantly different from your own.

Scenario: After the meeting, Javier approaches Debra and says, "I'd love to hear what managing a ranch is actually like."

Step 3: Practice Active Listening. Suspend judgment, listen fully, and allow your own assumptions to be questioned.

Scenario: Javier listens quietly as Debra describes the challenges of wildfire, drought, and regulation. He asks questions without interrupting.

Step 4: Embrace the "Fusion of Horizons". Allow understanding to arise from the encounter between your worldview and the other's—be open to synthesis rather than victory or defeat.

Scenario: Javier shares his own climate concerns. They realize both want sustainable futures, though their paths differ. Javier's assumptions about farming and resistance to change soften.

Step 5: Reflect and Integrate. Consider how the dialogue has altered, nuanced, or expanded your previous beliefs. Adapt your own position where appropriate and seek further dialogue to test and refine the new viewpoint.

Scenario: Javier writes afterward: "Farmers can be environmentalists too; their expertise is essential. My perspective is bigger after today."

A Real-World Example: You're convinced your community's traditions are inherently superior to others. *Application:*

- *Acknowledge Horizon:* Be honest about your upbringing and bias.
- *Seek Out Dialogue:* Attend an interfaith or intercultural community group meeting.
- *Practice Active Listening:* Listen deeply as a member from another tradition shares their perspective.
- *Fusion of Horizons:* Notice your perspective blending or expanding as you empathize with their experiences.
- *Reflect & Integrate:* Recognize new respect for difference and find value in diversity.

Here are a few tips: (1) Admit bias—write out your assumptions before conversation. (2) Set a goal to listen *without rebuttal*. (3) Ask open-ended questions ("Can you help me understand your view?"). (4) After each dialogue, write one thing you learned and one thing that challenged you. (5) Revisit your views and note what has shifted or expanded.

Process 3: Facilitating Transformational Learning

Perspective: Building on Mezirow's transformative learning theory, developmental psychology and adult learning propose that breaking through existing limits often comes through critical reflection on assumptions, usually triggered by a "disorienting dilemma" (Mezirow, 1991; Kegan, 1982).

Context for Scenario Example: Maria, a 47-year-old office manager, has always believed "You shouldn't talk about mental health in the workplace—it's unprofessional." She attends a company workshop on well-being.

Steps

Step 1: Encounter a Disorienting Dilemma. Experience (or purposely seek out) an event or contradiction that doesn't fit your current worldview.

Scenario: In the workshop, a colleague bravely shares their story about burnout. The positive response leads to improved team morale. Maria feels confused and conflicted.

Step 2: Critical Self-Examination. Reflect on the deep assumptions underlying your thinking and behavior; ask: "Why do I see it this way?" "What would it mean to think otherwise?"

Scenario: Maria asks herself, "Why do I think talking about stress is taboo? Is it just ingrained habit from my past jobs or family?"

Assumptions are the underlying beliefs or notions that shape our perceptions and influence our behaviors, often without us being consciously aware of them. To surface your own assumptions, reflect on a particular situation where you felt challenged or surprised by someone else's viewpoint. For example, imagine a workplace discussion where a colleague proposes a radically different approach to a project. Instead of immediately rejecting the idea, ask yourself *why you feel that way*. Are there underlying beliefs about how things should be done based on past experiences or traditions? Consider questions like, "What past experiences inform my reaction?" or "How might my upbringing shape my views on this issue?" By critically examining these assumptions, you can gain valuable insights into your own mindset and be open to alternative perspectives, ultimately facilitating transformational learning.

Step 3: Dialogue and Testing. Discuss your reflections with others who can offer challenges and support; actively consider alternative perspectives.

Scenario: At lunch, Maria asks the workshop speaker, "How did you decide to open up?" They discuss vulnerability and the benefits for the team.

Step 4: Reframe Core Assumptions. Deliberately revise or reconstruct your foundational beliefs or values in light of new insight.

Scenario: Maria decides, "Maybe professionalism and honest self-care can go together." She commits to trying something new.

Step 5: Experiment with New Roles/Actions. Try out behaviors, thoughts, and attitudes aligned with your transformed viewpoint.

Scenario: At the next staff meeting, Maria shares that she's using the company's wellness resources and encourages others to do the same.

Step 6: Build Competence and Confidence. Continue reflecting, learning, and adapting as the new frame becomes more natural, using challenges as growth opportunities.

Scenario: Team members approach her privately for support. Maria reflects proudly, "I helped change our culture for the better—something I never thought possible."

A Real-World Example: A lifelong meat-eater attends a documentary about animal rights. *Application:*

- *Disorienting Dilemma*: The film presents information that challenges their ethical beliefs.

- *Critical Self-Examination*: They reflect: "Why have I never questioned this before?"
- *Dialogue and Testing*: They reflect: "Why have I never questioned this before?"
- *Reframe Core Assumptions*: Decide, "Maybe my health and ethics can align differently."
- *Experiment with New Roles*: Decide, "Maybe my health and ethics can align differently."
- *Build Competence and Confidence*: Learn new recipes, join online communities, and reflect on the shift.

Here are a few tips: (1) When faced with discomforting information, pause to reflect rather than reject. (2) Journal your responses: "What values are being challenged?" (3) Find others who have made similar changes—share stories and resources. (4) Make a "small experiment" (e.g., a trial month or practicing a new behavior in a low-stakes setting). (5) Review progress regularly—celebrate growth and adapt as needed.

The Emotional Experience of Changing One's Beliefs

How does it feel when our beliefs shift? Belief systems shape identity, perception, and choice. When core beliefs shift—whether through dialogue, crisis, education, or personal reflection—the process is seldom emotionally neutral. Drawing on psychological literature and first-person reflection, this section explores the inner landscape of belief change, emphasizing the tension between discomfort and relief, and how both are integral to growth.

Human beliefs are not superficial add-ons; they are the frameworks through which we interpret the world, organize memory, make decisions, and define ourselves (Kelly, 1955; Kegan, 1982). When these frameworks shift—whether gradually or abruptly—the emotional consequences can be profound. To change a belief is, in a sense, to change one's world. How does this transformation *feel*? What emotional dynamics are at work in the "unsettling" and "opening" of the mind?

Shifting beliefs can cause considerable discomfort. This can be due to cognitive dissonance and uncertainty, loss and grief, or disorientation. Perhaps the first and most familiar feeling is *discomfort*, what, as introduced in Chapter 1, Leon Festinger (1957) called *cognitive dissonance*. When new information or experience contradicts our established beliefs, it creates mental tension. We may feel anxious, defensive, or even threatened. For example, someone raised with the conviction that their culture's values are universal might feel confusion, even frustration, when confronted by sincerely held, dramatically different values in another person. We dive deeper into cognitive dissonance in Chapter 11.

A second result is *loss and grief.* Beliefs are often tied to identity, family, and belonging. Questioning or losing them can evoke responses similar to mourning (Kegan, 1982). There is a sense of losing touch with old certainties, a world "disenchanted." For example, questioning a long-held religious or political identity may bring isolation from one's community or loved ones, triggering loneliness, guilt, or nostalgia.

A third result is *disorientation.* Without familiar signposts, previously "obvious" choices may become unclear. People report feeling adrift, uncertain of how to proceed, or who they are. For example, consider someone who has long identified as a staunch advocate for a particular political ideology. After engaging in discussions with diverse viewpoints and grappling with new perspectives, this individual begins to question the validity of their previously held beliefs. As they grapple with these conflicting ideas, they may find themselves feeling lost—not only do the ideas they once championed feel less solid, but the community and identity that came with those beliefs may also begin to fracture. This person might experience confusion about which values to embrace, how to engage in discourse with friends and family who still hold those views, or even what actions to take regarding political participation. The certainty that once guided their choices is replaced with ambiguity, leaving them in a state of emotional limbo as they navigate this new terrain of thought and identity.

When you are able to *release shifting beliefs*, there are a number of positive results including relief from dissonance, a sense of expansion and possibility, and reconnection and growth. Once the new perspective is genuinely integrated, the internal tension gives way to coherence. *Relief from dissonance* arrives—one's worldview and experience are once again in alignment (Festinger, 1957). For example, after wrestling with doubt, a person might experience clarity and calm, accepting a new understanding and letting go of an old internal conflict.

A second positive response is *a sense of expansion and possibility.* A new belief framework often opens space—a horizon beyond previous limits. People may describe feelings of liberation, curiosity, or excitement. *Example:* A formerly rigid thinker describes a "weight lifting" or becoming "open to the world in new ways."

A third positive response is *reconnection and growth*. New beliefs can bring reconnection—to new communities, sources of meaning, or aspects of self previously denied. For instance, imagine an individual who has spent years in a career that aligns with a customary, conventional path—one that reflects their upbringing and family expectations. After undergoing a significant personal transformation, they begin to question whether this path truly resonates with their authentic self. As they explore new interests and engage with different perspectives, they might be drawn to creative pursuits like art or entrepreneurship, which were once viewed as impractical. Through this exploration, they discover a vibrant community of like-minded individuals who share their passion and values. This newfound connection not only fosters a sense of belonging but also reignites a sense of purpose and inspiration. The individual feels pride in their bravery to embrace change and gratitude for the opportunity to connect with others who encourage their growth, ultimately leading to a richer, more fulfilling life aligned with their newly discovered beliefs.

Navigating the emotional landscape is an art form—no doubt one you have previously experienced. Nonetheless, the conscious shifting of beliefs can be quite demanding. First, you may experience ambivalence and mixed feelings of growth. These feelings are not separate stages but often coexist. Change can be messy. A person may mourn what is lost even as they celebrate what is gained; relief may be tinged with residual anxiety or guilt. This is when you may require support for integration. Research (Mezirow, 1991; Kegan, 1982) suggests that safe, supportive environments greatly help people process these feelings—allowing for mourning, dialogue, and the gradual weaving of new identities.

Belief shifts are emotionally rich events. They challenge and unsettle, but also release, expand, and renew. Growth inevitably carries pain—the discomfort of letting go—but also the possibility of joy and deeper connection as new turf is claimed. To shift belief is thus both a loss and a birth; both a crisis of meaning and its renewal.

Support During Belief Transition

Building on research and practice from psychology, education, and peer support, there are concrete approaches for supporting oneself and others during periods of belief transition. Here's a brief set for supporting oneself.

1. **Practice Self-Compassion**. *Acknowledge Normalcy:* Understand that discomfort, confusion, or grief are typical when core beliefs change (Neff,

2011). *Be Gentle:* Treat yourself with the care you'd offer a friend. Avoid self-judgment for feeling disoriented or "lost."

2. **Journal and Reflect**. *Record Emotions:* Writing about anxieties, hopes, losses, and gains helps clarify their sources and see patterns over time. *Notice Growth:* Regularly review journal entries to observe how your thinking and feelings evolve.

3. **Connect with Community.** *See Out Others:* Find peers, mentors, or communities who have experienced similar belief shifts. Online forums, support groups, or open-minded friends can provide reassurance and new perspective (Mezirow, 1991). *Share the Journey:* We are social beings. Articulating your experience often reduces isolation and validates the process of change

4. **Embrace Uncertainty.** *Allow Not-Knowing:* You don't have to resolve everything immediately. Uncertainty is a necessary and creative phase of growth. *Hold Space for Ambiguity:* Permit conflicting feelings to coexist without forcing a premature resolution.

5. **Mindfulness and Grounding.** *Stay Present:* Use grounding techniques (breathing, body scans, mindful walks) to stay rooted in the present moment if anxiety spikes. *Anchor in daily routines:* Maintain familiar habits to create stability as inner landscapes shift.

Just as you need support when moving through belief transition, others need your support when changing their beliefs. While recognition of this need may be difficult in a colleague, when there is open dialogue shifts in beliefs often move into the conversation. Here are some approaches to supporting others' shifts.

1. **Offer Nonjudgmental Listening**. *Be a Sounding Board:* Listen fully, without trying to "fix" or sway. Validate emotions, even those you don't share. *Ask Open, Clarifying Questions:* Encourage reflection ("What feels most unsettling?" "What new possibilities excite you?").

2. **Normalize the Experience**. *Share Gently:* Mention that belief change is a common part of growth; describe (or invite stories about) others who have navigated similar terrain.

3. **Provide Practical Support**. *Suggest Resources:* Recommend books, support groups, or educational material relevant to the transition. *Check in Regularly:* Sometimes, simply being present over time is most helpful.

4. **Model Patience and Acceptance.** *Resist Pressure to Hurry:* Everyone's pace is unique. Allow for pauses, reversals, or ambiguity in the other's

journey. *Accept Possible Outcomes:* Be open to the possibility that their new beliefs may challenge your own.

5. **Encourage Ongoing Reflection.** *Prompt Gentle Review:* Encourage journaling, conversation, or other forms of reflection to facilitate integration. *Celebrate Milestones:* Encourage journaling, conversation, or other forms of reflection to facilitate integration.

Belief transitions are often disorienting but also profoundly generative. Self-compassion, reflective practice, connection, and patient support are key tools for navigating such passages—whether for oneself or others. Growth, though sometimes painful, creates new ground for freedom and meaning.

Chapter 4
The Power of "Why"

There's power in asking just "Why?",
A question where truths often lie.
With purpose in sight,
Our paths can ignite
As we *Whysly* reach for the sky.

Many of us focused in the business world who have been a part of the movement into the knowledge age are familiar with the "Five Whys" problem-solving technique, which was developed by Sakichi Toyoda and is commonly used supporting Lean Manufacturing and Six Sigma methodologies. This exercise is designed to explore the root cause of a problem by repeatedly asking the question "Why?" The idea is that by drilling down into the layers of a problem, you can identify the underlying issue that needs to be addressed, rather than just treating the symptoms. Here are the steps:

Step 1: Identify the Problem. Start by clearly defining the problem you want to solve. This should be a specific issue that is impacting you or your organization.

Step 2: Ask "Why?" Once the problem is identified, ask "Why is this happening?" and note the answer. This first response might address a symptom or a surface-level cause.

Step 3: Repeat. For each subsequent answer, ask "Why?" again. Continue this process, typically for five iterations, although it may vary based on the complexity of the problem.

Step 4: Analyze the Response. As you continue to ask "why," you will begin to peel back the layers of the issue. The goal is to reach a root cause that gives a clearer understanding of the problem.

Step 5: Develop Solutions. Once the root cause is identified, brainstorm potential solutions to address it. This might involve making changes to processes, improving communication, or addressing specific issues that led to the problem.

There are numerous benefits to this methodology. First, its *simplicity*. The technique is straightforward and doesn't require advanced statistical methods or complicated analysis. Second, its *focus on root causes*. It encourages you to go beyond surface-level issues and focuses on systemic problems. Third, it *engages teams*. It can encourage collaboration and input from multiple stakeholders who might have insight into different aspects of a problem. And fourth, it *promotes continuous improvement*. Regularly applying the Five Whys can foster an organizational culture that values solving root causes rather than just addressing symptoms.

Of course, on reflection, there are several limitations as well. First, what we are getting is a *superficial analysis*. If not conducted thoughtfully, the Five Whys can lead to oversimplified conclusions. Second, there is a *dependency on knowledge*. The effectiveness relies on the knowledge and honesty of the participants in addressing the real issues. We all know that an individual's knowledge is limited (thus the need for knowledge sharing and working in teams and communities), and we also recognize that knowledge is context-sensitive and situation-dependent, always being created and re-created for the immediate issues at hand, all of which affect the quality of knowledge. Third, *the answer is not always found in "five whys."* Sometimes, the root cause may surface in fewer questions … or need more.

Still, the Five Whys exercise is a powerful tool for diagnosing problems and finding solutions at their source, contributing to more effective problem-solving practices in various fields.

The Larger Concept of "Why"

The "power of why" refers to the fundamental importance of understanding the motivations behind our actions, beliefs, and goals. At its core, asking "why" allows individuals and organizations to clarify their purpose and direction. Simon Sinek (2009) popularized this idea with his concept of the "Golden Circle," which illustrates that the most successful leaders and organizations begin by identifying and communicating their core "why"—the reason they exist beyond just making a profit.

The *Golden Circle model* emphasizes the importance of purpose and aligns an organization or individual's actions with their core beliefs. It consists of three concentric circles, each representing a different level of communication and understanding. The innermost circle is the **Why**, signifying the *core belief or purpose*. This is the fundamental reason an organization or individual exists—beyond just making money. It's about why you do what you do and what inspires you.

The middle circle represents the **How**, the *process or values* that explain how you go about achieving your "why." This includes the unique selling points or approaches that set you apart from others in the same field. The outer circle encompasses the **What,** the *products or services* that you offer. It answers the question of what you do—tangible outputs that can be easily identified.

Sinek argues that *most people and organizations* communicate from the outside in, starting with "what" they do and "how" they do it before addressing "why." However, this approach can lead to a feeling of disillusionment and alienation in an individual and, in an organization, to a lack of emotional connection with customers or team members. By flipping the approach and starting with "why," individuals have increased motivation and enhanced fulfillment, and organizations can create a stronger, more inspiring connection between their purpose and their audience.

A prime business example is Apple. Instead of just saying they sell computers (the "what"), they communicate their belief in challenging the status quo (the "why"). They create products that are not only innovative but also embody their values, which resonates deeply with their customers. Sinek (2009) also examples Martin Luther King, Jr., highlighting how Dr. King communicated his dream and vision for equality (the "why") rather than just focusing on specific actions or legislation (the "what"). This approach inspired millions and garnered widespread support for civil rights.

However, the application of this model goes beyond the business context. Individuals can use the model to clarify their own purpose and values, guiding their life choices and actions toward what truly matters to them. By emphasizing the importance of "why," Sinek highlights that understanding core beliefs can lead to more authentic and impactful engagement, whether in business, leadership, or personal development. And leaders who articulate their "why" can generate loyalty and motivation within their teams, fostering a culture that aligns with shared values rather than just tasks.

Relating to this Book's "Why" Question

Our question is: "Why Can't I see What's Right in Front of Me?" As simple as that question appears, in the complex world of humans and humanity it is going to be incredibly difficult to simplify and highly context-sensitive and situation-dependent.

[First Exercise] Let's assume you are struggling to recognize an important opportunity in your life—perhaps a job offer, a relationship, or a creative project. Using the Five Whys, here's how you might work through it. *Problem Statement*: "I can't see the opportunity in front of me."

1. *Why can't I see the opportunity in front of me?* Because I feel overwhelmed by my current responsibilities and commitments.

2. *Why do I feel overwhelmed by my current responsibilities?* Because I haven't set boundaries, and I'm trying to manage too many projects at once.
3. *Why haven't I set boundaries?* Because I fear disappointing others and want to be seen as helpful and reliable.
4. *Why do I fear disappointing others?* Because I derive a lot of my self-worth from being perceived as valuable to others.
5. *Why do I derive my self-worth from others' perceptions?* Because I haven't defined my own values and what makes me feel fulfilled outside of external validation.

Several insights have emerged from this exercise. First, re *boundaries*. You may need to assess your commitments and learn to say no to protect your time and mental space, allowing you to see opportunities more clearly. Your actions might include: (1) Create a plan to establish boundaries that protect your time and energy. (2) Identify where you can cut back on commitments to focus on what is essential.

Second, *self-reflection*. Recognizing that your self-worth is tied to external perceptions points to a need for introspection. This realization might prompt you to engage in activities that foster internal validation—such as journaling, practicing self-compassion, or setting personal goals. As you use the Five Whys process, it will often come down to your own perception and understanding of your "self," the *who you are*. Your actions might include incorporating mindfulness techniques or practices to help you stay present, which can assist you in recognizing opportunities as they arise.

Third, *identifying values*. The final "why" reveals a crucial step—defining your values and what fulfillment looks like for you. With a clearer sense of purpose, you may find it easier to recognize the opportunities that align with your true self. Your actions might include engaging supportive trusted friends or mentors who can help you explore your thoughts and perceptions, providing perspective and encouragement in your journey.

[Second Exercise] Looking at the chaotic environment that is part of our day-to-day living and seeking to develop a larger understanding, let's reverse the question and reflect on it from a less situationally-dependent specific viewpoint. Let's ask: "*Why do I see what I see?*"

1. *Why do I see what I see?* Because my perception is influenced by my personal experiences and beliefs.

2. *Why are my perceptions influenced by my personal experiences and beliefs?* Because my past experiences shape my understanding of the world and how I interpret new information.
3. *Why do my past experiences shape my understanding?* Because they create a framework of references that I use to evaluate situations and make judgments.
4. *Why do I use a framework of references to evaluate situations?* Because it helps me make sense of the world and navigate complexities more easily.
5. *Why do I need to make sense of the world and navigate complexities?* Because I seek to find meaning, safety, and certainty in my life, which informs my decision-making and interactions with others.

Through the Five Whys methodology we have uncovered the following insights. First, re *personal experience*. Our perceptions are deeply rooted in our past experiences and beliefs, indicating that understanding ourselves better can improve our perception. Second, re our *framework for evaluation*. We rely on a cognitive framework to interpret information, showing that this framework can vary significantly from person to person, leading to different perspectives on the same situation. For some people this may include emotional indicators.

Third, our *desire for meaning*. Our quest for meaning and certainty in life is fundamental to how we perceive and respond to the world around us. This reflects our inherent need to feel grounded and understood. Again, we see that by the fifth "why" it becomes very personal, in this case reflecting the inherent human need to not only understand ourselves but to be understood by others.

This exercise highlights that perceptions are not merely observations; they are filtered through personal narratives, past experiences, and cognitive frameworks. Recognizing this can encourage you to explore different viewpoints and increase your openness to new experiences, ultimately enriching your understanding of the world.

[Third Exercise] While certainly closely related to our second exercise, let's now address our core question. *Problem Statement:* I can't see what's right in front of me.

1. *Why can't I see what's right in front of me?* Because it doesn't fit into my mental models or paradigms.
2. *Why doesn't it fit into my mental models for paradigms?* Because my mental models and paradigms are shaped by my past experiences and beliefs, which define what I consider possible or valuable.

3. *Why are my experiences and beliefs shaping my mental models and my paradigms?* Because they create a lens through which I view the world, influencing and bounding my interpretations of situations and decisions.
4. *Why do these beliefs and experiences impact my interpretations?* Because they establish expectations and assumptions (predictions) that guide my understanding and responses, often limiting my openness to new ideas.
5. *Why do I hold onto those expectations and assumptions?* Because they provide a sense of stability and security, making it easier to navigate life's complexities, although potentially at the cost of missing new opportunities.

The insights derived from this exercise include. First, *paradigmatic limitations.* Understanding that your inability to see what's in front of you is linked to your established paradigm highlights the importance of challenging and expanding your existing beliefs. Second, the *influence of past experiences.* It underscores how past experiences and deeply ingrained beliefs can create blind spots, making it difficult to recognize new opportunities or perspectives. This insight is critical since we live in a world that is continuously changing—and challenging—our assumptions such that old ways of thinking and doing rapidly become insufficient to survival and success.

Third, the *echo chamber effect.* Understanding that you may gravitate toward information that confirms your beliefs shines a light on the need for openness and curiosity to broaden your perspective. And fourth, the *need for stability.* Again, digging into the core of "self" of who we are, the desire for stability can lead to a resistance to change, indicating that embracing flexibility and openness can enhance your perception and decision-making.

This exercise illustrates that the rigidity of a personal paradigm which can significantly impact your perception and awareness. It emphasizes the need to actively seek out diverse viewpoints and challenge your assumptions to enrich your comprehension of the world. By recognizing the limitations of your current learning and understanding, you can consciously work to expand your horizons.

The "Why" of Being Human

Understanding your "why" is a vital component of the human journey of discovery. It serves as a foundational element that shapes our identify, guides our decisions, and influences our relationships. In a larger sense, it plays a critical role in our philosophy and social structure. Taking a meta-perspective of why "why" is important, particularly from the viewpoint of humanity and applying a consilience approach, let's explore this from the perspectives of existential inquiry, cultural development, scientific progress, motivation and agency, connection and empathy, and the evolution of consciousness, all elements related to our core question.

Existential Inquiry. *Search for Meaning:* Throughout history, humanity has grappled with existential questions about purpose and existence. Asking "why" allows individuals and societies to explore their place in the universe, fostering a deeper understanding of life's complexities. *Philosophical Foundations:* Philosophers from Socrates to Nietzsche have emphasized the importance of questioning as a means to achieve wisdom. The pursuit of "why" challenges assumptions and prompts critical thinking, pushing humanity towards greater intellectual development.

Cultural Development. *Human Narratives:* Every culture has its creation myths, moral philosophies, and belief systems that address "why." These narratives provide context and structure, helping to bind communities together and establish collective identities. Similarly, each individual creates their own personal narrative, contiguously linking together the spatial and temporal events that are most significant to who they are. *Evolution of Values:* Societal norms and values often arise from collective inquiries into "why." As societies evolve, these values are tested and redefined through discourse, leading to innovations in ethics, governance, and social interaction.

Scientific Progress. *Curiosity and Exploration:* The questioning of "why" is at the heart of scientific inquiry. From the Scientific Revolution to modern advancements, asking "why" drives exploration and understanding of the natural world, leading to technological breakthroughs and improved quality of life. *Understanding Reality:* Scientific inquiry seeks to uncover the laws of nature and the underlying principles of existence. This pursuit not only enriches knowledge but also enhances humanity's ability to make informed decisions and adapt to changing environments.

Motivation and Agency. *Driving Action:* Understanding "why" provides motivation for individuals and communities to act. It fuels passions, inspires change, and galvanizes movements. Social justice, environmental advocacy, and humanitarian efforts often stem from individuals asking "why" about injustices and inequalities. *Empowerment:* When people understand their motivations and the reasons behind their actions, they feel more empowered to pursue their goals and effect change. This sense of agency contributes to personal and societal growth.

Connection and Empathy. *Building Relationships:* Asking "why" promotes deeper connections between individuals. By understanding each other's motivations, fears, and values, people can cultivate empathy and compassion, which are essential for fostering strong communities. *Navigating Differences:* In an increasingly globalized world, understanding the diverse "whys" that drive different cultures and individuals is crucial for peaceful coexistence and collaboration. It encourages dialogue, reduces conflict, and promotes understanding.

Evolution of Consciousness. *Self-Reflection:* The act of questioning encourages self-reflection and introspection, which are essential for personal growth. By examining "why" we think and act as we do, individuals can develop a more profound awareness of their beliefs, values, and behaviors. *Collective Consciousness:* On a societal level, collective questioning about "why" can lead to shifts in awareness and consciousness, motivating societies to evolve and adapt to new challenges, such as climate change, inequality, and technological disruption

In essence, the importance of "why" transcends individual understanding and resonates at the core of humanity's quest for meaning, connection, and progress. It drives existential inquiry, fosters cultural development, propels scientific advancement, and shapes our relationships and social structures. By continuously exploring "why," humanity can grow more conscious, compassionate, and capable of navigating the complexities of existence.

Exploring Our Individual "Why"

Viewing the search for "why" as a fundamental reason for being and living is a profound perspective. It touches on the essence of human existence and the intrinsic need to find purpose and significance in our lives. While there will be some overlap with the why of being human, here are some reflections on how this search for "why" shapes our individual understanding of who we are, directly impacting what we "see" and what we don't "see".

1. **Defining Purpose**. *Core of Existence:* The quest for "why" is central to defining what gives our lives meaning. Understanding our purpose can anchor us during turbulent times, providing direction and motivation. *Individual Significance:* Each person's "why" will differ, influenced by passion, values, and experiences. This diversity and individuality is what contributes to the richness of the human experience, emphasizing that each life holds unique significance.

Relationship to our core question: Without a clarified purpose, our capacity to *recognize*, *know*, or *understand* what truly matters is diminished—even when it is present and accessible. When we lack a clear sense of purpose, it's not just that we don't visually notice what's around us; we also fail to *recognize* meaning and opportunities in our own lives. Without purpose as a lens, significant truths about ourselves, our relationships, or our environment may remain unacknowledged or misunderstood. *Example:* Someone unclear about their own values may not *recognize* a career or relationship that truly aligns with who they are, even if it's directly before them.

2. **Connection to Humanity**. *Shared Journey:* The search for purpose unites people across cultures and generations. Regardless of background, the desire to understand one's place in the world is a shared human trait. This creates a sense of solidarity and connection among people. *Empathy and Understanding:* As you explore your own "why," you become more attuned to the struggles and journeys of others. This fosters empathy and compassion, aligning you with a broader human narrative.

   Relationship to our core question: Our evolving sense of connection determines who and what we are able to *"see"* in our social world, sometimes making us oblivious to available relationships or shared experiences Our search for place and meaning among others shapes not just what we physically perceive, but also who and what we *acknowledge*, *understand*, and *relate to*. Blind spots may persist in our empathy, recognition of others' struggles, or even our own need for connection, simply because we are unconsciously filtering what (and who) we acknowledge as relevant to our "why". *Example:* Someone preoccupied with belonging may fail to *recognize* another's outreach or the subtle ways they are already included.

3. **Motivation for Action**. *Driving Force:* A clear sense of "why" can propel you into action. It motivates you to pursue your passions, engage in meaningful work, and contribute positively to your community or the world at large. *Finding Joy in Struggles:* When your actions are guided by a strong purpose, challenges and setbacks may be viewed as necessary steps toward fulfilling that purpose rather than just obstacles. This perspective can enhance resilience and determination.

   Relationship to our core question: When motivation is unfocused or confined, potential actions, resources, or even our own capabilities may go unseen, unrecognized, or unacknowledged. With an unclear or overly rigid "why," we may miss not just visible opportunities but also fail to *recognize* when is the right moment to act, or *know* that we possess the ability or permission to move forward. We can be unaware of our potential or alternatives that lie within reach. *Example:* Someone so focused on a traditional route to success may not *know* that a non-traditional opportunity—right in front of them—could bring greater fulfillment.

4. **Navigating Life's Challenges**. *Framework for Decision-Making:* A well-defined "why" acts as a compass in navigating decisions and challenges. When faced with adversity, returning to your "why" can clarify your choices and help you stay true to your ideals. *Creating Meaning from Adversity:* The search for "why" can lead to a transformative experience, especially in times of hardship. Finding meaning in suffering or challenges can contribute to personal growth and a deeper understanding of life.

   Relationship to our core question: Struggles can cloud not just our vision but our insight, making it difficult to "see" value, meaning, or growth—right in the midst of our experience. In times of adversity, not only do we miss cues in our environment, but we also might fail to *understand* the meaning in those experiences, or to *know* how to draw strength or learning from them. We could be blind to our own resilience, helpful insights, or avenues for growth that are present but unrecognized. *Example:* Someone going through hardship may not *realize* their growth until much later, or fail to *understand* the support being offered by loved ones.

5. **Legacy and Impact**. *Contributing to the Collective Good:* Understanding your "why" often leads to a desire to leave a positive impact on the world. This sense of contribution can provide a profound sense of fulfillment, enhancing your sense of interconnectedness with humanity. *Shaping Future Generations:* By living authentically and in alignment with your "why," you inspire others to explore their own purposes and contribute to a legacy of purpose-driven living.

   Relationship to our core question: Our legacy is often built from small acts and present choices we do not *recognize* as significant until much later—they are "right in front of us" yet go unacknowledged. If we do not reflect on our "why," we may fail to *notice* the ripple effects of our everyday actions, or lack awareness of the immediate impact we have on others. Conversely, living in the distant future may blind us to the influence we have in the present, meaning we don't *know* the significance of small, daily choices. *Example:* A teacher might not *realize* the lasting impact of a kind word to a student in the present moment.

In all these areas, "seeing" encompasses not only physical perception, but also *knowing, understanding, recognizing,* and *acknowledging* the people, opportunities, and meanings that surround us. The clarity or obscurity of our "why" determines what we are able to truly become aware of in our lives.

Articulating Your Own "Why"

Uncovering and articulating your own "why" is a deeply personal and reflective process. It involves introspection, exploration, and sometimes even experimentation. Below are steps and strategies to guide you in discovering your "why".

Step 1: Self-Reflection. *Journaling:* Spend time writing about your experiences, feelings, values, and aspirations. Prompt yourself with open-ended questions like: What moments have made me feel the most fulfilled? What activities energize me and make me lose track of time? What do I truly care about, and why? *Identify Core Values:* Reflect on what values resonate with you. These can be honesty, compassion, joy, growth, service, adventure, etc. Write down the values that matter most and think about how they influence your decisions and actions.

Step 2: Explore Your Passions. *Engage in Activities You Love:* Participate in various activities, hobbies, or volunteering. Pay attention to what excites you and what feels meaningful. This can provide insight into what you might want to pursue further. *Notice Patterns:* Look for recurring themes in your interests. If there are specific topics, causes, or fields that consistently draw you in, they may point towards your "why."

Step 3: Reflect on Significant Experiences. *Life Events:* Identify key moments in your life that have shaped who you are. Consider both positive and negative experiences. What lessons did you learn from these moments? How have they influenced your perspective? *Inspirational Figures:* Think about people you admire and the qualities they embody. What is it about their lives or actions that resonates with you? What values do they represent that you wish to emulate?

Step 4: Ask Deep Questions. *The Five Whys:* Think about people you admire and the qualities they embody. What is it about their lives or actions that resonates with you? What values do they represent that you wish to emulate?

Step 5: Experiment and Adapt. *Try New Things:* Don't be afraid to step out of your comfort zone. Engage in new experiences that intrigue you, whether it's traveling to a new place, learning a new skill, or meeting new people. Each experience can provide insight into what aligns with your "why." *Iterate:* Your "why" might evolve over time as you gain new experiences and perspectives. Be open to revisiting and refining your understanding of it regularly.

Step 6: Seek Feedback. *Engage Others:* Discuss your reflections and insights with trusted friends, mentors, or coaches. They can provide valuable perspectives on your strengths, passions, and potential blind spots. *Active Listening:* Pay attention to how others perceive you and what they think you are passionate about. Sometimes, external perspectives can highlight aspects you may not fully recognize in yourself.

Step 7: Draft Your Statement. *Articulate Your "Why":* Once you have a clearer understanding of your motivations and values, try to articulate your "why" in a concise statement. This statement should reflect your core beliefs and what drives you. *Test and Refine:* Share this statement with trusted individuals to get feedback. Be willing to edit and refine it until it feels authentic and resonates with you.

Step 8: Live Your "Why". *Align Actions with Purpose:* Once you've articulated your "why," aim to align your daily actions, decisions, and goals with this purpose. This alignment can lead to greater satisfaction and fulfillment.

Let's reflect on these steps through a story.

The Community Garden Project (A Story)

Context: The central character in this scenario is someone named Alex, who goes on a journey that ultimately leads to initiating a community garden project, addressing food insecurity and fostering community connections. This scenario exemplifies how the process of uncovering one's "why" can lead to impactful actions that bring about positive change in the community.

[*Step 1: Self*-Reflection] Alex begins by journaling about his life experiences. He writes about feeling unfulfilled in their corporate job and reminiscing about his childhood when he spent time gardening with his grandmother. Through these reflections, Alex realizes he finds joy and peace in nurturing plants and connecting with nature.

[*Step 2: Explore Your* Passions] Motivated to reconnect with his childhood love for gardening, Alex starts attending local workshops on sustainable gardening and permaculture. He discovers a passion for growing food and learns how community gardens can alleviate food insecurity while promoting environmental sustainability and social interaction.

[Step 3: Reflect on Significant Experiences] Alex reflects on a time when he volunteered at a local food bank, witnessing firsthand the struggles of families facing food insecurity. This experience ignites a desire to make a difference and creates a sense of urgency in addressing hunger in their community. He also thinks about his grandmother's teachings about the importance of sharing and community.

[*Step 4: Ask Deep Questions (The Five Whys)*]. Alex starts questioning his motivations for wanting to make a change:

- *Why do I want to address food insecurity?* Because I believe everyone deserves access to healthy food.
- *Why is that important to me?* Because I experienced food scarcity in my childhood and saw how it affected my family.
- *Why does that impact me deeply?* It has shaped my understanding of community and the role of support systems.
- *Why do I want to foster community?* Because I believe that connectedness can lead to resilience and joy.
- *Why do I care about connectedness?* I've seen how isolation can harm mental health, and I want to create spaces that bring people together.

[*Step 5: Experiment and* Adapt] With newfound clarity, Alex visits several community gardens to understand their dynamics. He engages with community members, learning what works and what doesn't. Alex also experiments with growing various vegetables and herbs in a small section of his backyard, sharing the harvest with neighbors and seeking their input. This feedback helps refine his vision for a larger project.

[*Step 6: Seek* Feedback] Alex organizes a small neighborhood gathering to discuss the idea of creating a community garden. He presents his vision and goals, emphasizing the importance of having a collective space for cultivating food and relationships. Neighbors express enthusiasm and share their own stories of wanting to contribute, providing feedback and ideas that enhance the project concept.

[*Step 7: Draft Your* Statement] After gathering insights and understanding their motivations, Alex crafts a personal "why" statement: "My purpose is to cultivate a sense of community and resilience by creating a garden where individuals can grow food, forge connections, and support each other in nurturing a healthier, more sustainable future."

[*Step 8: Live Your "Why"*] With his "why" clearly articulated, Alex mobilizes a group of neighbors to establish a community garden. They secure a vacant lot through collaboration with local organizations and city officials.

Over the next few months, they design, plant, and care for the garden together. Workshops on gardening, cooking, and nutrition become regular events, further enriching the community.

Through this meaningful journey, Alex transforms personal reflections and passions into communal action, addressing food insecurity and fostering relationships. The community garden becomes a symbol of hope and connection, demonstrating how one person's quest for "why" can lead to greater goodness in the world. This initiative not only provides fresh produce but also strengthens community bonds, supports mental well-being, and inspires collective stewardship of the environment.

The "Why" and "Who" Entanglement

The relationship between understanding your personal "why" and defining "who" you are is profound and intertwined, with the "who" I choose to be very much driving the answers to the "why" I am. This is a personal relationship, with the personal "why" (purpose, motivation, or meaning) deeply intertwined with one's "who" (identity, self-concept, or the person one chooses to be). Our understanding of purpose shapes and is shaped by the sense we have of our own identity, with each continually informing and redefined the other in the iterative journey of life.

The search for purpose is rarely an abstract, disconnected endeavor. Instead, it unfolds within the context of our evolving sense of self. The stories we tell ourselves about who we are—our values, strengths, cultural narratives, and even aspirations—act as filters that influence what kind of purpose feels authentic and compelling. For instance, a person who sees themselves as a healer may find their "why" rooted in acts of service or care, whereas someone who identifies as an innovator may be most driven by curiosity, risk, and change. In this way, "who" we choose to be becomes the soil from which our sense of "why" grows.

Conversely, clarifying and living from a sense of "why" powerfully shapes our identity over time. When we engage with a meaningful purpose, our actions, decisions, and relationships begin to align with that purpose, slowly transforming our sense of self. The repeated pursuit of meaning—through career, love, art, service, or adventure—helps us integrate these experiences into our core identity. In this way, purpose both emerges from and recursively forms our chosen identity. As Viktor Frankl (2006) wrote in his seminal work *Man's Search for Meaning*, it is the will to meaning that animates human life, and in choosing our attitudes and responses to life's challenges, we simultaneously discover both our purpose and our truest self. The important role of "will" is addressed in Chapter 16.

Ultimately, the boundaries between "who" and "why" are porous. Identity is not fixed, nor is purpose static; each illuminates and expands the other. We become who we are by seeking and living our purpose, and in the becoming, we refine and sometimes even discover new "whys" for our existence. This dynamic interplay is, in many ways, at the heart of the human journey. With this critical entanglement in mind, the author offers the personal exercise that helped her discover and become her authentic self.

Defining Your Authentic Self (An Exercise)

Here's a detail guide on how to carry out this structured exercise for discovering and defining your authentic self. This exercise serves as a pathway to uncovering your authentic self by actively integrating traits you admire into your life. By consistently living according to these characteristics (both setting intent and ensuring attention to these characteristics), you become more aligned with your true self, ultimately leading you to deeper insights about your existence and purpose. Over time, this reflective practice not only transforms your mindset but can profoundly impact your relationships and contributions to the world.

Step 1: Identify Three People Who Inspire You. *Choose Three People:* Think of three people you admire, love, and respect. These can be public figures, mentors, friends, or family members. The key is to select people whose qualities resonate deeply with you. *Write Their Names:* Jot down their names and a brief description of your relationship with each (e.g., "My grandmother – she showed me kindness and resilience").

Author Example: My three were my father, my mentor (a musician; I was an opera singer in my younger years) and then-Sister Teresa.

Step 2: Explore Why You Admire Them. *Reflect on Your Feelings:* For each person, ask yourself why you admire them. Consider their actions, behaviors, values, and the impact they've had on you. *Prompt Questions:* What qualities do they embody? How do they influence their surroundings? What do they do that evokes feelings of admiration or respect? *Identify Characteristics:* Write down two standout characteristics for each person, aiming for a total of six traits (two from each person you have identified).

Author Example (who did this exercise in her early 30's):

(1) My father, Demetrious Vasileios (Basil) Sioris. His qualities were many. Until I met my 30-year life partner, David Hughes Bennet, recently passed, there was no one who held a candle to my father! He arrived in the U.S. in 1911 at age 12 with a 7th grade education and no English. Yet he achieved the American dream, owning his own home, starting his own restaurant, sending doweries to 6 of his 7 sisters so they could marry, wedding a young French-ancestry American, and adopting and raising two abandoned children when he had none of his own. There was a continuous stream of people who came to him for his help and advice, which was always available. He was **wise** and **kind**.

(2) My mentor, Dr. Norman Söreng Wright, out of the old French school having studied with Maurice Ravel, composer of orchestral and piano music, including *Boléro*, associated with impressionism, and Marcel Dupré, renowned 20th-century French organist and composer with incredible virtuosity and improvisation skills, famous in the world of sacred and organ music. Dr. Wright, whether naturally or from years of study I do not know, definitely displayed those same skills on the organ. He had strong integrity but what I loved about him is that with that integrity he emanated a flow of life energy which caught you up in in its music with moments of pure joy and love. So, the characteristics had to be **integrity** and this wonderful **flow state of joy and love**.

(3) Then-Sister, soon to become "Mother", Teresa. Mother Teresa had unwavering compassion, humility, and dedication to serving the poorest and most vulnerable. With extraordinary empathy and kindness, she offered comfort and care to those who were sick, abandoned, or dying, seeing dignity in every human life regardless of circumstance. Her deep sense of selflessness and sacrifice was evident in her willingness to live among those she served and to put the needs of others before her own. Guided by a profound faith, she inspired countless others with her relentless commitment to charity, simplicity, and unconditional love, leaving a lasting legacy of hope and humanity. Two traits were hard to choose. The first was her **compassion**. The second was chosen because of my own acknowledged weaknesses (at the time very much an empath and having not developed the left side of my brain) and that was her **persistence**. She never waivered in her mission. Life requires that strength.

Step 3: Analyze the Characteristics. *List the Traits:* Now you have a total of six characteristics or traits. Write them down in a list. *Evaluate Their*

Importance: Reflect on each trait. Why is each important to you? How do they align with the vision of who you want to be?

From the author: To accomplish this step I had to look from the outside-in, that is, imagine my conscious self looking at this person who I was choosing to be. What did they tell me about this person? Selecting wisdom, kindness, integrity, the flow of love and joy, compassion, and persistence as your guiding characteristics revealed a deep commitment to both personal growth and the well-being of those around you. This combination suggested a striving to be thoughtful and discerning (wisdom), making choices rooted in understanding and perspective. The emphasis on kindness and compassion highlighted a desire to be caring and supportive, consistently sensitive to the needs and struggles of others. Integrity spoke to a dedication to authenticity and honesty, ensuring that actions aligned with values, while persistence showed strength, a determination to continue progressing and serving others, even when faced with difficulties. The flow of love and joy—foundations for a life well-lived and often the most difficult to achieve—expressed both an openness to positivity and the wish to spread warmth and hope while moving through life.

Together, these qualities create a harmonious and resilient foundation for relating to the world. Wisdom guides choices; kindness and compassion shape interactions; integrity grounds character; persistence fuels actions and efforts; and the flow of love and joy enriches presence and existence. Such a combination represents someone who aspires to understand and uplift others, but also someone who remains true to oneself and presses on with hope and courage, regardless of circumstances. These qualities, working in concert, reflect an intention to live with meaning, to be a force for good, and to leave a gentle, enduring impact on the world around you. The difficult question posed: *Was this possible?*

Step 4: Integration into Daily Life. *Set Intentions.* Do not attempt to embrace all six characteristics at the same time; rather, weave them into your everyday life slowly, giving time and thought to exactly how your thoughts, feelings and actions reflect these characteristics you have chosen to be. Pick the strongest two traits from the list to start with (you can rotate in the others later). These will be your primary focus. *Daily Reflection:* Before making conscious decisions or taking actions on those decision, ask yourself: "How do I embody these traits in this moment?" For example, if one of your traits is "kindness," consider how you can act kindly in your interactions today. *Journaling:* This is critical. Keep a daily or weekly journal to reflect on how you embodied these traits. Note specific instances where you succeeded and instances where you struggled, jotting down how you think and feel, and how you might better reflect the WHO you have chosen to be. Because this WHO is NOT the future WHO, but the current WHO you are because of your choices. **Think it. Feel it. Act on it.**

From the author: While the first week or two worked well, the process became more difficult as I fell back into old patterns that included conflict avoidance. Still—persistence embraced—I kept my focus and continued the important work of being who I chose to be. The journal became ever more important.

Step 5: Develop Self-Awareness. *Reflect on Consistency:* Over time, pay attention to how consistently you embody these traits. Are there patterns in situations where you feel aligned with your chosen traits? Are there situations where you find it challenging to act in alignment with them

From the author: Yes to both questions. It was necessary to track my daily thoughts and feelings right along with my decisions and actions, ensuring that these chosen characteristics were consistent across all of these. There was much work to be done, but an incredible sense of satisfaction when, one day, the flow happened with a positive sense of accomplishment. One day. But then there were more days and my understanding of the authentic me I chose to be expanded.

Step 6: Reflective Review. *Schedule Regular Reviews:* (Weekly) Review your reflections and identify successes and areas for growth. Take note of circumstances that tested your adherence to your traits. (Monthly) Assess your progress in internalizing these traits. Have any new qualities emerged? Do you still honor the traits you chose? *Rotate traits:* Once you feel comfortable with the initial traits, consider integrating the remaining four into your life.

Step 7: Explore Your Purpose. *Ask Yourself "Why Am I Here?":* After investing time in becoming/ensure you are the person you wish to be, reflect on the broader question of your existence. Consider how your newfound traits contribute to your purpose. What roles do you feel called to play in your relationships, community, or the world? *Journaling Prompts:* What unique contributions do I wish to make? How do my traits align with my passions and desired? In what way do I want to impact others or my community?

From the author: My life has taken many twists and turns. So much to share, but that is another book, perhaps. What is important in terms of this exercise, is that for the next ten years (on and off) I reviewed and ensured consistency with my authentic self, intentionally ensuring this was who I am, and that my decisions and actions reflected that "who," although we're all human, and we all make mistakes. Still, there was continuous improvement. THEN, ten years later, I was working in Yokosuka, Japan, as an employee with the U.S. Navy, serving at the time as editor of the base newspaper, when I was personally called to spend a day with Mother Teresa on her first, and to my knowledge only, visit to a military base. It was a life-changing experience (Bennet, 2018; Bennet & Scott, 2021), which made a believer of me. **Where you set your intent and what you focus your attention upon is what comes into your life**.

She leaned towards me, perhaps a product of the bent form, and tilted her leathery face upward so that her eyes could take in my face, could follow my expression. Her mouth was slightly, knowingly, curled upward at the edges, laughing at the moment, and her eyes ... and her eyes ... I don't have the words. (Photos by Alex Bennet)

Chapter 5
The Convergence of Consilience

In the field where disciplines blend,
A consilient path we defend.
With convergence in sight,
We *Whysly* unite,
Finding truths as perspectives extend.

Consilience is a convergent process. Let's spend a few minutes exploring what is meant by that statement, then apply it to the core question we are addressing in this book: *Why can we not see what is right in front of us?*

The concepts of convergence and consilience are indeed intertwined, especially in terms of how they relate to multidisciplinary approaches in understanding complex systems (humans and the organizations of which they are a part) or phenomena. Convergence generally refers to the process where different entities or fields come together to reach similar conclusions or solutions. This can happen in various contexts, such as scientific research, technology development, or even within social sciences. Convergence emphasizes the idea that despite different starting points, methods, or disciplines, there can be a unifying direction or understanding that emerges over time.

Conversely, consilience is the principle that evidence from independent, unrelated sources can converge to achieve strong conclusions. It was popularized by E.O. Wilson, who argued that findings from different disciplines can overlap and support a unified theory. Consilience is often seen in the context of synthesizing knowledge from various fields—like biology, sociology, and economics—to create a cohesive understanding of a complex issue.

When we use the concepts convergence and consilience, there is considerable overlap in terms of interdisciplinary collaboration, integration of knowledge, complexity in understanding, and theoretical development. Related to interdisciplinary collaboration, both concepts emphasize the importance of collaboration between fields. That makes sense. As researchers from different areas work together, their unique perspectives and methodologies converge towards a common goal, which fosters consilience.

In terms of the integration of knowledge, convergence typically leads to a more integrated understanding of a topic. When researchers apply methods from various disciplines to a single problem, they can create a more comprehensive framework that reflects the nuances of the subject matter. This mirrors the goal of consilience, where the ultimate aim is to create an overarching narrative supported by diverse evidences.

Related to complexity in understanding, as the fields converge, so too do the complexities of the subjects they study. Consilience helps to unpack these complexities by validating findings across different domains, ultimately leading to a more robust and multifaceted understanding.

In theoretical development, the convergence of different disciplinary approaches often leads to the development of new theories that are enriched by the insights of diverse fields—a hallmark of consilience. For instance, in addressing climate change, insights from environmental science, economics, and social policy must converge to create effective strategies.

Viewing consilience as a convergent process highlights the dynamic interactions between disciplines that advance our collective understanding of complex issues. Both concepts encourage collaboration and integration, and reflect the evolving nature of knowledge in a rapidly changing world. So, how does this understanding of the overlap between these two concepts help us address our core question?

Applying a Convergent Approach to Our Consilient Exploration

Let's have some fun and create a step-by-step convergent approach to exploring the idea of why we don't see what is right in front of us, with examples from various fields such as neuroscience, psychology, sociology and spirituality. This will help us determine how to organize the chapters in the rest of the book, while also serving as an introduction to some of the ideas presented in later chapters.

Step 1: Define the Core Question. *Action:* Clearly articulate the central inquiry—what specific aspects of not seeing what's right in front of us do you want to explore?

Response: For purposes of this exercise, let's add some specificity to our core question. How about: "What cognitive, emotional, and social factors prevent individuals from recognizing significant personal truths or societal realities?"

Step 2: Literature Review Across Disciplines. *Action:* Conduct an extensive literature review in each targeted field to gather existing research and theories related to your core question.

Response: The Mountain Quest Institute is blessed with a 40,000+ book research library. Thus, the in-depth chapters related to the various fields are rich with information. Limiting this exercise to four primary fields for this exercise, we might choose to focus on the following:

- *Neuroscience*: Look into how the brain processes visual information and the role of intention and attention in perception.
- *Psychology*: Investigate cognitive biases, such as confirmation bias, that might cloud judgment and perception.
- *Sociology*: Examine social constructs that shape collective awareness or ignorance (e.g., groupthink).
- *Spirituality*: Explore insights from spiritual traditions about awareness, perception, and inner "seeing."

Step 3: Synthesize Findings. *Action:* Identify common themes, patterns, or contradictions in the reviewed literature.

Response: We find that neuroscience indicates limited perceptual capacity (i.e., selective attention), psychology reveals cognitive biases, sociology discusses the influence of social groups, and spirituality offers a perspective on inner awakening. This representative synthesis highlights a multidimensional understanding of perception.

Step 4: Conduct Interdisciplinary Interviews. *Action:* Involve experts from each field to gain insights and fresh perspectives that can enrich your understanding.

Interview (1) a *neuroscientist* to discuss how visual processing can lead to missed details.

Interviewer: Thank you for joining me today. Can you explain how our brain processes what we see, and why we might miss important details in our environment?

Expert: Absolutely. Our brain relies on a selective attention mechanism, which means it filters out excess sensory information to focus on what it deems important. This is why we can miss details that are right in our line of sight—our brain prioritizes certain stimuli over others based on its past experiences, expectations, and learned biases. For example, when people are engrossed in a task, they may literally fail to see things that are unusual or unexpected in their environment. This phenomenon is often illustrated through studies on "inattentional blindness."

Interviewer: How do cognitive overload and distractions influence our perception?

Expert: Cognitive overload can significantly impact our ability to process information. In today's world, we're bombarded with stimuli, and our brains can become overwhelmed, leading to a reduced capacity to notice important details. Distractions also play a key role. When our focus is divided, we may overlook significant cues or signals that would otherwise catch our attention.

Interview (2) with a *psychologist* to explore emotional factors affecting perception.

Interviewer: It's great to have you here. Could you elaborate on how cognitive biases influence our perception and lead us to overlook important truths?

Expert: Certainly. Cognitive biases, such as confirmation bias, affect how we perceive information. People tend to attend to information that confirms their existing beliefs and to ignore or dismiss evidence that contradicts them. This can prevent them from seeing uncomfortable truths—like relationship problems or societal injustices—simply because acknowledging these would challenge their worldview.

Interviewer: Can you share how emotional states might influence what we notice in our lives?

Expert: Emotions can act as a filter for perception. When someone is experiencing high levels of anxiety or depression, their focus may narrow to negative aspects of their lives, causing them to miss positives or solutions. Conversely, positive emotions can widen one's perspective, making them more open to new experiences and insights. This emotional state plays a critical role in shaping attention and awareness.

Interview (3) with a *sociologist* to discuss societal pressure to conform, influencing what people notice.

Interviewer: Thank you for participating. How do social dynamics and group behavior contribute to our collective blindness to certain issues?

Expert: Social dynamics can create a concept known as groupthink, where cohesiveness in a group leads to irrational decision-making. Members may suppress dissenting viewpoints to maintain harmony, resulting in a collective ignorance toward problems that need addressing. Additionally, social norms can dictate what people pay attention to or discuss, leading to a societal blindness toward issues that don't align with those norms, such as systemic inequalities.

Interviewer: How does the concept of social capital relate to perception?

Expert: Social capital refers to the networks and relationships within a community. People with strong social ties might share a common perspective that can cloud their ability to see divergent viewpoints or alternative realities. This tendency can lead to a homogenized perception of issues, where differing experiences—such as those of marginalized groups—are overlooked or ignored.

Interview (4) with a *spiritual teacher* to reflect on the practices that enhance awareness.

Interviewer: Thank you for joining the discussion. How do spiritual practices contribute to enhancing awareness and improving perception?

Expert: Spiritual practices such as meditation and mindfulness center around cultivating present-moment awareness. By regularly engaging in these practices, individuals can learn to observe their thoughts and feelings without attachment, allowing them to see beyond surface-level concerns. This heightened awareness can illuminate truths that they may have overlooked in their daily lives.

Interviewer: In what ways do spiritual teachings address the phenomenon of "blind spots"?

Expert: Many spiritual traditions emphasize the idea of ego and attachment as barriers to perceiving reality clearly. For example, teachings often convey that our identification with material success or social status can prevent us from seeing our true selves or the needs of others. By recognizing and transcending these attachments, individuals can open themselves to deeper truths and realities that have been obscured by their personal and societal conditioning.

***Synthesis*:** These sample interviews reflect how each field contributes valuable insights into the overarching question of perception. Neuroscience offers an understanding of cognitive mechanisms, psychology highlights emotional and cognitive biases, sociology explores the influence of social dynamics, and spirituality addresses the importance of awareness practices. Each expert provides a unique lens to examine why we might not see what is right in front of us, enriching the discourse across disciplines.

Step 5: Develop Case Studies. *Action:* Use real-world examples or hypothetical scenarios to illustrate how people miss what's right in front of them across different contexts.

Examples: (Neuroscience) A scenario where a person overlooks details in a complex visual environment due to distractions. (Psychology) A character who stays in a toxic relationship, ignoring signs of its dysfunction due to attachment. (Sociology) A community that collectively overlooks social injustices due to groupthink. (Spirituality) A person who experiences moments of clarity and suddenly "sees" their truth through meditation.

Case Study (1): The Neuroscience of Inattentional Blindness. *Scenario:* Sarah is a busy graphic designer who frequently works on complex projects. One day, her team conducts a brainstorming session to develop a new marketing campaign. During the meeting, they decide to put together a slide deck, and Sarah is responsible for visualizing their ideas. *Analysis:* As Sarah works on her laptop, she becomes so focused on aligning her visuals with the team's ideas that she fails to notice a colleague walking in and waving to her. Neuroimaging studies suggest that Sarah's focus on her task resulted in selective attention, leading to "inattentional blindness." This case illustrates how our cognitive processing can lead us to overlook stimuli in our environment, even when they are in plain sight (Simons, 1999). *Conclusion:* To mitigate this, Sarah might benefit from periodic breaks or exercises that emphasize mindfulness, helping her to broaden her attention and notice important interactions and details beyond her immediate task.

Case Study (2): Psychological Blind Spots and Relationships Dynamics. *Scenario:* Mark and Lisa have been in a relationship for several years. While they are generally happy together, Mark starts to feel increasingly dissatisfied but fails to recognize the extent of his discontent. He frequently rationalizes away minor conflicts, attributing them to stress from work rather than addressing them directly. *Analysis:* Psychology researchers note that Mark's cognitive biases, such as confirmation bias and the emotional toll of avoidance, contribute to his inability to see the relationship's underlying issues. His emotional state, compounded by the fear of confrontation, creates a blind spot, making it difficult for him to acknowledge that significant changes might be necessary for both partners' happiness (Seligman & Csikszentmihalyi, 2000). *Conclusion:* Encouraging open communication and possibly engaging a therapist could help Mark and Lisa navigate these hidden issues, obtaining insights that they might have overlooked due to avoidance and emotional barriers.

Case Study (3): Sociology and Groupthink in Workplace Culture. *Scenario:* A software development company, InnovateTech, has a strong team-oriented culture. The team is comprised of individuals who frequently agree with each other's ideas. When a significant flaw is discovered in a product before launch, members of the team hesitate to voice their concerns due to fear of upsetting the group's harmony. *Analysis:* This case exemplifies the sociological phenomenon of groupthink. The pressure to conform stifles dissenting opinions and critical thinking. Researchers explain that such dynamics can lead to collective blindness towards problems that are apparent to outsiders. Team members fail to see the severity of the flaw because they prioritize group cohesion over transparency and critical discussion (Janis, 1972). *Conclusion:* Implementing a system that encourages anonymous feedback or appointing a "devil's advocate" could help facilitate a culture of open dialogue, enabling team members to address potential issues before they escalate.

Case Study (4): Spiritual Awakening and Inner Sight. *Scenario:* David is a successful corporate executive who has dedicated his life to climbing the corporate ladder. He finds himself increasingly anxious and unfulfilled despite his external accomplishments. One day, he decides to attend a meditation retreat recommended by a friend. *Analysis:* During the retreat, David engages in mindfulness practices that foster self-reflection and inner awareness. This experience allows him to confront his attachment to success and status—elements that he previously believed defined him. Spiritual teachings emphasize the liberation that comes from recognizing one's true self beyond societal expectations. David begins to see the aspects of his life that he had previously ignored, such as his genuine passions and the importance of relationships (Kabat-Zinn, 2005). *Conclusion:* Through the lens of spirituality, David's journey underscores the value of introspection and mindfulness in revealing truths that may have been obscured by societal conditioning. His awakening signifies the ability to shift perspective and embrace alternative ways of understanding life beyond material success.

Summary: These case studies collectively demonstrate how different fields can provide insights into the reasons behind our perceptual blind spots. They illustrate real-world scenarios that highlight the complexity of human perception and the factors that inhibit our ability to recognize critical truths.

Step 6: Identify Converging Insights. *Action:* Look for insights that emerge from the integration of findings and case studies across disciplines.

Converging insights from interviews and case studies. Based on the interviews and case studies, several converging insights emerge regarding why we often fail to see what is right in front of us.

Insight (1): *Selective Attention.* Both neuroscience and psychology reveal that our brain's selective attention mechanism can lead to inattentional blindness. This filtering process allows us to focus on specific details while ignoring others, impacting our awareness of critical information.

Insight (2): *Cognitive and Emotional Biases.* Psychological factors such as cognitive biases (e.g., confirmation bias) and emotional states (e.g., anxiety, attachment) significantly influence our perceptions. These biases can create blind spots, causing individuals to overlook essential truths in personal relationships and broader contexts.

Insight (3): *Social Influence and Group Dynamics.* Sociological insights emphasize that social dynamics, including groupthink, can lead to collective blindness to important issues. The pressure to conform within groups can suppress dissenting opinions, leading to the overlooking of critical problems.

Insight (4): *Awareness and Spiritual Practices:* Spirituality offers a pathway to overcoming perceptual blind spots through mindfulness and introspection. Practices that enhance self-awareness enable individuals to recognize and confront underlying beliefs and attachments that obscure their reality.

Insight (5): *Integration across Disciplines.* All these insights point toward an interconnected framework, suggesting that understanding human perception requires an interdisciplinary approach. Each perspective enhances our comprehension of how cognitive, emotional, social, and spiritual dimensions interplay in shaping awareness.

Step 7: Create a Synthesis Framework. *Action:* Develop a coherent framework that ties together insights from all disciplines under your central theme.

Scenario Example: Create a model that illustrates how attention, bias, social context, and spirituality interact, leading to "blind spots" in awareness. This framework can include interconnected components such as attention processing (neuroscience), cognitive distortions (psychology), social dynamics (sociology), and awakening processes (spirituality).

Framework Title: The Convergence of Awareness Framework (CAF)

Components of the Framework:

(1) ***Cognitive Filters*** (neuroscience). *Key Insight:* Attention mechanisms filter sensory information—leading to selective perception. *Action Point:* Recognize and train the brain to expand attention through mindfulness exercises.

(2) ***Emotional Awareness*** (psychology). *Key Insight:* Emotional states and cognitive biases shape how we interpret information. *Action Point:* Encourage emotional literacy and awareness to help individuals confront subconscious biases.

(3) ***Social Context*** (sociology). *Key Insight:* Social dynamics and group norms influence collective perception and blind spots.

(4) ***Spiritual Insight*** (spirituality). *Key Insight:* Social dynamics and group norms influence collective perception and blind spots. *Action Point:* Social dynamics and group norms influence collective perception and blind spots.

Framework Interactions:

- ***Cognitive Filters ↔ Emotional Awareness****:* Understanding how emotions influence attention can help individuals become more aware of their cognitive biases.
- ***Social Context ↔ Cognitive Filters****:* Group dynamics can either enhance or inhibit cognitive filtering. Encouraging diverse perspectives within groups can broaden attention and awareness.
- ***Spiritual Insight ↔ Emotional Awareness****:* Spiritual practices can foster emotional regulation, providing individuals avenues to address emotional states that cloud judgment.

Implementation in real-world contexts might include workshops and training sessions, and mindfulness programs. Schools could include curriculum integration, staff training, and after-school programs. Workplaces would include wellness programs, quiet spaces, and leadership buy-in. Healthcare would include patient education, staff training, and Integration into Treatment Plans. Community programs would include public workshops and mindfulness groups.

Implementing mindfulness in various real-world contexts requires careful planning and adaptation to meet the specific needs of the target audience. By actively involving stakeholders, providing adequate training, and maintaining a supportive environment, mindfulness practices can lead to significant improvements in mental well-being and overall life satisfaction

Step 8: Write and Iterate. *Action:* Draft your book, integrating the framework and case studies. Focus on how the converging insights inform your core question, and seek feedback from experts in the respective fields to refine your ideas.

Scenario: As you write, let the content drive the format, weaving together narratives, case studies, and expert insights, showing the complexities of perception. In addition to scenarios, you might share personal anecdotes, professional case studies, and empirical research to bring your message to life.

Step 9: Engage with Readers. *Action:* Share drafts or summaries with a diverse audience, including people from each of your focus disciplines, to elicit feedback and foster dialogue.

Scenario: Share with colleges for reflection and thoughts. Host workshops—virtual or face-to-face. This can enhance your understanding and provide new perspectives.

Step 10: Finalize and Publish. *Action:* Incorporate feedback, make necessary adjustments, and prepare your manuscript for publication.

Scenario: After thorough revisions and input from various stakeholders, publish your book, ensuring it addresses the interdisciplinary connections and insights that explain why we often don't see what's right in front of us.

After Thoughts

This introductory convergent process begins to focus our core idea through initial discussion and examples, but, as the steps unfold, also demonstrates the rich interplay of diverse disciplines as we grapple with a complex question. The very nature of our inquiry—into why we often cannot see what lies plainly before us—resists being bounded or organized by field or discipline. Instead, it calls for us to identify and integrate theories, concepts, and approaches from psychology, philosophy, neuroscience, spirituality, and beyond.

But first, let's develop a foundation to build upon, a basic understanding of some of the primary fields we will pull from and their central contributions to our focal point.

Chapter 6
Conscious Experience

In the realm where consciousness reigns,
Awareness breaks through its old chains.
With minds open wide,
We *Whysly* decide
To see past habitual lanes.

The concept of "field" is larger than the concept of "discipline". A field refers to a general area of study, work or interest while a discipline is a distinct area of study or branch of knowledge, often with its own methods, theories, and terminology. Disciplines tend to have established academic departments, scholarly journals, and formalized methods of inquiry.

A field may include several related disciplines or subfields. For example, the field of "science" includes disciplines like biology, chemistry, and physics. In our area of focus, the field of cognitive science includes psychology, neuroscience, AI, philosophy, linguistics and anthropology, as well as others. As an analogy, we could think of a field as a forest, and a discipline as a particular type of tree in that forest.

There are also "studies" that intersect several disciplines. For example, the study of consciousness is a multifaceted topic that intersects multiple disciplines, but is most prominently situated within the following fields: philosophy, cognitive science, neuroscience, psychology, phenomenology, and cultural studies.

What is Consciousness?

Talk about coming up with unanswerable questions! While having read literally hundreds of books which forward a potential answer to this question—which generally includes awareness of self and the other, most often with the caveat that there is no agreement on this definition—I tend to lean towards the "descriptions" forwarded by my recently departed partner and spouse, David Hughes Bennet, who was both a nuclear physicist and a neuroscientist, and my also recently departed spiritual mentor, Jo Dunning. We build on the idea of awareness of self and the other, with the description of consciousness as a process, a sequential set of ideas, thoughts, images, feelings and perceptions and an understanding of the connections and relationships among them. Added to that is the acknowledgement that consciousness is the sum total of who we are, what we believe, how we act and the things we do, so it's all in our actions, thoughts and words. This works well for someone who has focused personally and professionally on the importance of knowledge (the capacity to take effective action) for some 40-50 years!

When developing his early experiential learning model—upon which the Kolb model and the expanded Bennet model (ICALS) are built—Dewey noted that "the principle of continuity of experience means that every experience both takes up something from those which have gone before and modifies in some way the quality of those which come after" (Kolb, 1984, p. 27; Bennet, Bennet, & Turner, 2022). Today, we recognize that the "experience" of life is contiguous rather than continuous, that is, the created "story" of our life, which emphasizes the physical and spatial proximity and adjacency between and among events, is concerned with the direct connections or relationships between things that are positioned next to each other in time *and* space (Bennet, 2025). So, with this understanding, we can describe consciousness as a contiguous rather than continuous process.

The above description points out the importance of at least a partial understanding of consciousness to answer why we can't see what is right in front of us. It is our "consciousness," however you perceive it, that is "seeing" (or not seeing) what is right in front of us. Since the conscious experience is directly relative to understanding our question, let's briefly look at the aspects of consciousness focused upon in the various fields making large contributions to researching our core question.

[*Philosophy*] Philosophers have long grappled with questions of consciousness, including the nature of subjective experience, the mind-body problem, and issues surrounding personal identity. Key philosophical questions include what it means to be conscious, how consciousness relates to the physical brain, and whether there are different types or levels of consciousness.

[*Cognitive Science*] This interdisciplinary field studies the processes of the mind, including perception, memory, decision-making, and consciousness itself. Cognitive scientists investigate how consciousness arises from neural processes and how it influences cognition and behavior.

[*Neuroscience*] Neuroscience seeks to understand the biological underpinnings of consciousness by studying brain activity and structure. Researchers in this field use various methods—including brain imaging techniques—to explore which neural correlates are associated with conscious experience.

[*Psychology*] In psychology, consciousness is explored in terms of awareness, attention, and the relationship between conscious and unconscious processes. Researchers examine how consciousness affects behavior and mental processes, and how alterations in consciousness can occur (e.g., through sleep, drugs, or certain psychological disorders).

Though often associated with philosophy, *phenomenology* also plays a role in the study of consciousness as it emphasizes the subjective experience of being conscious and how individuals perceive and interpret their experiences. In addition, researchers in *cultural studies* examine how concepts of consciousness vary across cultures and how cultural practices influence the understanding of conscious experience.

Anthropology examines the cultural contexts of consciousness and perception, investigating how different societies interpret experiences and what they may prioritize or overlook. *Sociology* investigates how social structures and interactions influence perception and awareness, particularly concerning collective norms and group dynamics. *Religious studies* explore various spiritual and religious perspectives on consciousness, including altered states of consciousness that can affect perception and awareness. *Psychiatry* examines consciousness in the context of mental health disorders, understanding how conditions like schizophrenia or dissociative disorders can drastically alter perception and recognition.

As can be seen, these fields expand the interdisciplinary approach to understanding consciousness, emphasizing that the study of *perception and awareness*—that which enables us to see what is or is not right in front of us—is multifaceted and influenced by numerous factors, including cultural, social, and biological dimensions.

Physiological and Cognitive Mechanisms of Awareness

The journey from "raw" sensory input to conscious awareness is far from straightforward. Conscious awareness is commonly misunderstood as a passive, comprehensive "recording" of reality, in which our senses deliver a faithful representation of our environment to a central observer. Yet, mounting evidence from neuroscience, psychology, and cognitive science demonstrates a different story: *awareness is not passive, but actively constructed and highly selective.* Only a small fraction of sensory input reaches conscious awareness, with most filtered, ignored, or even altered before it emerges into what we call experience (Dehaene, 2014; Baars, 1988). The very processes that support our intelligence—efficiency, prediction, and goal-orientation—ensure that we miss, misinterpret, or reinterpret much that is "right in front of us."

This constructed nature of awareness means our conscious experience is shaped by both biological imperatives and learned models. As Hawkins (2021) notes, our "intelligent" brains are designed less to record every detail and more to make the best possible inferences in a complex, changing environment. Understanding the mechanisms by which this happens is key to unpacking why we so often fail to see the obvious, and why consciousness itself remains such a profound mystery.

The *physical process* of becoming aware begins with the transduction of sensory signals—light, sound, touch, etc.—into nerve impulses. But what's striking is how much *filtering* and *transformation* these signals undergo before they ever reach consciousness.

At the lowest level, sensory organs and peripheral nerves already discard data. For example, our retinas respond only to a small slice of the electromagnetic spectrum, and our brains adapt to constant stimuli (such as touch or sound), ensuring only changes stand out (Kandel, Schwartz, & Jessell, 2013).

Once signals reach the brain, the *thalamus* acts as a major relay station and filter, directing select information to primary sensory cortices and inhibiting irrelevant inputs (Crick, 1994). This early editing reflects evolutionary priorities—what is unfamiliar, emotionally salient, or surprising gets promoted, while the familiar fades into the background.

In the **cerebral cortex**, information flows through hierarchical networks (Hubel & Wiesel, 1962), where it is not just "read" but interpreted. It is compared to prior experiences (the concept of associative patterning), and grouped into patterns (Bennet, Bennet, & Turner, 2022). Many signals are processed entirely outside awareness, in parallel, by specialized modules (Milner & Goodale, 2006), such as those for recognizing faces, motion, or language.

This networked, modular brain (a key theme in Hawkins work on the intelligent brain) means that awareness emerges *as a product of both distributed and integrated activity*, not from a single "center" or localized structure. As a result, what we become aware of depends as much on what our *brain expects* (intention, prediction) and what modules communicate, as on what is objectively presented (but subjectively processed).

Central to our understanding of awareness is the role of the brain's neural connectivity and the dynamic interplay between different regions. The concept of neural integration highlights how disparate sets of information from sensory inputs are synthesized, with feedback loops allowing for rapid adjustments based on context and prior experiences. This integration facilitates not just the perception of stimuli but the formulation of predictions about future events, enabling the brain to navigate complex environments efficiently. Such mechanisms underline the brain's capacity for adaptive behavior, where conscious awareness emerges from both the accumulation of sensory data and the active modulation of that data through cognitive processes. By exploring the intricate architecture of neural networks responsible for these functions, we can better appreciate the sophisticated nature of consciousness and the selective spotlight it casts on our experiences.

Conscious Access

Neural transmission is not enough for conscious access: *awareness requires attention*. Selective attention acts as a spotlight, enhancing relevant information and suppressing the rest (Posner, 1980).

The **Global Workspace Theory** offers a robust model for understanding how local neural activity becomes conscious awareness. The theory posits that many processes in the brain operate unconsciously and in parallel, but only when information is "broadcast" to a widely distributed frontoparietal network—the "global workspace"—does it become available to consciousness, verbal report, and voluntary control (Baars, 1988; Dehaene & Changeux, 2011). Barrs likens this to a theater spotlight: only what is illuminated can be seen by the "audience" of our cognitive systems. Dehaene and Changeux use brain imaging to show that perceived (vs. missed) stimuli trigger a broad, sudden, and sustained pattern of activity—"ignition"—across multiple brain regions.

This selectivity is crucial. Without it, the system would be flooded by overwhelming amounts of data. But it also means that when attention is elsewhere, even dramatic events can go unnoticed—an effect vividly demonstrated in inattentional blindness and change blindness research (Simons & Chabris, 1999; Mack & Rock, 1998). (See Chapter 9 on inattentional blindness and Chapter 7 for an in-depth treatment of attention and intention.)

Prediction

The brain's strategy for handling limited resources is not just to filter and spotlight, but *to actively predict what it expects to encounter*. **Predictive coding** posits that perception is fundamentally inferential, that is, the brain constructs internal models of the world, generates predictions about incoming sensory data, and only updates these models when prediction errors occur (Friston, 2005; Clark, 2013).

Most sensory input is "explained away" by these models, and only the *unexpected* or *unpredicted* is selected for further processing—or for *conscious awareness* (Hawkins, 2021). This is not a flaw, but an evolutionary necessity to enable quick, adaptive functioning in an environment of overwhelming complexity. In our evolution, it was never advantageous to perceive all incoming stimuli; instead, it was crucial to attend only to those elements that might indicate opportunity or threat, or challenge our survival-relevant expectations (Friston, 2005; Feldman, 2013).

As a result, our conscious experience is less an unfiltered, objective window onto reality and more a finely tuned summary of what the brain predicts will matter—a contiguous experience rather than a continuous experience. When a new stimulus matches prediction, it is often ignored or relegated to the background; when it deviates significantly, error signals are generated, capturing attention and updating our mental models (Clark, 2013). (For a description of mental models, see Chapter 13, Bits and Pieces, under Cognitive Science.) In this way, our brains work as active sculptors, continually refining a "best guess" model of the world—always based on prior beliefs, context, and goals.

This predictive scaffolding means that what is "right in front of us" is frequently discounted or rendered invisible if it doesn't fit the brain's current expectation set. Take, for example, the famous "gorilla experiment" (Simons & Chabris, 1999). When individuals focused on counting basketball passes, the appearance of a person in a gorilla suit became virtually invisible—not because the eyes did not register it, but because the predictive models of "what is relevant during this task" did not allocate representational resources to the gorilla. The brain efficiently "explained away" the mismatch as noise, or did not surface it to consciousness at all.

Jeff Hawkins' "Thousand Brains" theory (2021) further emphasizes that each cortical column constructs its own model of the world, producing a multitude of predictions based on both sensorimotor input and prior learning. Integrating these multiple viewpoints—sometimes in the face of conflicting information—enables flexible perception and fast action, but also opens the door to various forms of cognitive blindness: if the ensemble agrees on a faulty prediction, disconfirming data is more likely to be ignored—our collective "guess" simply overlays the unexpected, cementing our inability to see what is actually present.

Thus, predictive coding elegantly explains both the efficiency and fallibility of human awareness. It is an adaptive tradeoff, maximizing survival while inevitably *leaving us blind to the many truths we do not expect.*

Hawkins' Model and the Modular Approach

Hawkins "Thousand Brains" theory adds elegant depth to our understanding of distributed and modular intelligence—and helps explain why missing the obvious may be more the rule than the exception. According to Hawkins (2021), the neocortex is composed of hundreds of thousands of similar modules which he calls "cortical columns," each building its own partial model of the world. These models are constructed via sensorimotor learning. Each module interprets input in terms of spatial relationships, sequences, and reference frames. Perception then emerges from consensus among these modules, using cross-talk to reinforce or inhibit signals based on learned expectations.

While this distributed system yields powerful pattern recognition and allows us to learn the structure of the world from a multitude of perspectives, it also means that *collective error* can occur. If the majority of modules predict a stable, unremarkable environment, novel or anomalous input may remain beneath the threshold of conscious awareness—regardless of its actual salience. This is the neural substrate for why multiple observers can miss the same anomaly, and why deeply ingrained expectations called "priors" guide not only what we see, but what we notice at all (Hawkins, 2021; Clark, 2013).

Hawkins also emphasizes that the flexibility of the cortical modules—continually updated according to sensory-motor experiences—means that what is "obvious" to one person in a given context may be invisible to another. Context-specific learning shapes the reference frames by which the world is mapped and understood (Kandel, Schwartz, & Jessell, 2013). Thus, awareness is both a shared biological inheritance and a unique, personal construct—a further explanation for why blindness to the evident can profoundly differ across individuals or cultures.

Implications

Bringing together insights from predictive coding and modular intelligence, the answer to "why we can't see what's right in front of us" reveals both an adaptive strength and a profound limitation. By prioritizing efficiency, relevance, and prediction, the brain enables swift navigation of a complex world—but at the cost of filtering out the unexpected, the ambiguous, and the evolutionarily "irrelevant." Our awareness is the byproduct of layered, dynamic selection—sculpted by learning, attention, goal states, and cultural context.

There are personal and collection actions that can be taken to expand our capacity for awareness. First, *reflection and metacognition*. Recognizing the limits of our constructed consciousness is a call to cultivate metacognitive habits—actively questioning our assumptions, seeking novelty, and designing environments that encourage broader attention. Mindfulness practices, for example, can "de-automatize" habitual filtering and surface previously overlooked details (Lutz et al., 2008).

Second, *operationalizing awareness*. In organizations, science, or education, structured feedback, deliberate perspective-taking, and cross-disciplinary dialogue can help challenge entrenched models and surface the "invisible gorillas" in collaborative contexts.

Third, *technological considerations*. As AI and digital tools increasingly shape what we notice and value, understanding these brain-based blind spots is vital to designing systems that compensate for, rather than exploit, our attentional vulnerabilities (Dehaene, 2014).

The mechanisms that give rise to consciousness and construct our sense of NOW are not defects but ingenious adaptations. Yet, in situations of rapid change, uncertainty, or ambiguity, these very mechanisms can prevent us from seeing what is present and possible. The invitation is to embrace both humility and curiosity—knowing our seeing is always partial, and that with effort, inquiry, and creative challenge, we can shift both individual and collective awareness to include more of what truly matters (Bennet, 2025b).

Let's create a few exercises.

Exercise (1): Awareness Expansion—Seeing Beyond Habit

Here is a step-by-step awareness-expanding exercise, grounded in cognitive science, mindfulness research, and practical psychology. The goal of this exercise is to guide you through a sequence that gently disrupts habitual predictive processing, invites sensory and attentional openness, and cultivates metacognitive reflection. This process can begin to help us "see what is right in front of us," including what we normally filter out.

Scenario Context: Imagine you are sitting at your kitchen table in the morning. The exercise below is demonstrated with specific inner dialogue and actions to make the process intuitive.

Steps

Step 1: Set Intention (Meta-Attentional Priming). *Action:* Begin by stating aloud or silently: "For the next few minutes, I intend to notice what usually escapes my attention." *Purpose:* This primes your brain to shift from automatic prediction to exploratory mode, gently lowering the threshold for novelty and surprise (Lutz et al., 2008).

Scenario Example: You sit at your kitchen table, coffee mug in hand. You think to yourself: "For the next few minutes, I intend to notice what usually escapes my attention in this space."

Step 2: Ground in Sensation (Sensory Reset). *Actions:* Sit or stand comfortably. Take a slow, deliberate breath. Close your eyes for a few seconds and attend to sounds, physical sensations of touch, temperature, and your bodily posture. Open your eyes, allowing your gaze to softly widen; rather than focusing on one object, attend to the entire field of vision. *Purpose:* Starts to disrupt your brain's expectation-driven processing by focusing on direct, raw sensation, not categories or labels.

Scenario Example: You put down your phone, take a deep breath, and close your eyes. You notice: (*Visual*) The texture of the chair under your legs. (*Smell*) The aroma of coffee. (*Sound*) The faint hum of the refrigerator. (*Feel*) The gentle air against your skin. You open your eyes, softening your gaze to take in the whole room, not just your table.

Step 3: Guided Observation—Novelty Seeking. *Action:* Pick an ordinary space around you (your room, office, outdoors). Look for something you have never noticed before—a detail, pattern, object, or subtle change. If nothing comes up right away, look more slowly and deliberately, as if you were a visitor seeing this place for the first time. *Tip:* Give yourself at least 60–90 seconds. Allow discomfort or impatience—these are signs your brain is pushing against its predictive habits (Clark, 2013; Lutz et al., 2008).

Scenario Example: You begin scanning the room as if it's new, resisting the urge to label things automatically. You notice the obvious—placemats, salt shaker, kettle—but look further. A small, brown stain on the far wall—never noticed before. You stare a bit longer: "Was that always there?" You spot a tiny crack along the edge of a tile, the way the morning sun glints off the chrome toaster. You realize the kitchen clock is actually ticking—a sound you've tuned out for years

Step 4: Attentional Shifting—Multi-Sensory Exploration. *Action: Choose a single "new" object or detail.* Allow yourself to "describe" it silently, as if explaining to someone who cannot see it—this recruits language centers, deepening conscious encoding (Baars, 1988).

- *Visual:* Notice its color, shape, texture, shadows.
- *Touch* (if possible): What does it feel like?
- *Hearing*: Does it make any sound?
- *Smell/Taste* (if appropriate): Are these senses engaged?

Scenario Example: You fix your attention on the small, brown stain. (*Visual*) You notice the irregular shape, a darker edge that fades to light. (*Touch*) You get up, touch the wall gently—It's smooth except the edge, which feels tacky. (*Sound*) You hold your ear near—no sound, but in that stillness, the clock's tick sounds louder. (*Smell*) You sniff—no odor; you wonder if it's coffee that splashed weeks ago. (*Visual*) You also notice the curve of steam rising from your mug, the subtleties you usually miss.

Step 5: Mindful Noting—Label Your Experience. Action: Without judgement, silently note: What am I *expecting* to see/feel? What actually *is* here? What surprises me, even a little? This activates "metacognitive awareness," spotlighting the gap between internal expectation and external fact (Lutz et al., 2008).

Scenario Example: You think to yourself: (1) "I expected the wall to be blank—I never thought to look for stains." (2) "I assumed my kitchen was silent, but it is alive with tiny sounds." (3) "The stain surprised me, and also how shiny the toaster actually is in sunlight."

Step 6: Reflect and Record. *Action:* Take 1–2 minutes to jot down or voice-record: What did you notice that was previously invisible? How did it feel to look/listen this way? In what other contexts might you have "missed the obvious" due to habit?

Scenario Example: You grab your journal or dictate to your phone: (1) "I never realized how much detail I miss here daily. The wall stain, the ticking clock, the way sunlight shifts on metal. I see how easily I move through life on autopilot, missing the textures and stories around me." (2) "It feels unusual at first—almost uncomfortable, like slowing down in rush hour. But after a minute, things seem richer and…slower."

Step 7: Integration—Set a Micro-Intention for the Day. *Action:* As you return to your routine, choose a specific action: "Today, I'll pause and deliberately notice one new thing during [routine activity: walking, eating, driving, etc.]." This intentional "pause" builds the habit of questioning prediction and broadening attention, reinforcing neural pathways for expanded awareness (Dehaene, 2014).

Scenario Example: As you finish your coffee, you say to yourself: "Next time I take a work break, I'll pause and notice at least one new detail in my office before starting up again." You realize: you've shifted into a more open, curious mind—primed to catch what habitually slips by.

Optional Extension: Try this exercise with a friend or group. Compare "missed" details and reflect on how each person's predictive models shape unique blind spots.

Through this process, your predictive brain—usually blinded by habit—relaxes its grip. You reclaim access to "the obvious," bringing to light familiar details made invisible by routine. Over time, repeating such exercises in different settings can fundamentally sharpen your awareness, creativity, and capacity for presence.

Exercise (2): Memory vs. Reality—"Seeing Your Street" Challenge

This exercise focuses on making the limits of habitual perception concrete—and on experiencing, in a direct and embodied way, how much we "fill in" from memory and expectation rather than true observation. Here's a step-by-step expansion, blending mindfulness, predictive coding awareness, and creative practice.

Steps:

Step 1: Choose and Visualize. Think of a street you know well—somewhere you walk often, perhaps because of a favorite café, bookstore, or your daily commute. Close your eyes and visualize: Where do you usually start? What do you pass by on the way to your destination? What do the buildings, sidewalks, trees, shops, and people look like? Are there particular sights, sounds, or smells you associate with this place?

Step 2: Draw or Map from Memory. (1) Take a blank sheet of paper and sketch the street (it's okay if you're "not an artist"—stick figures, simple shapes, and labels are perfect). (2) Mark out the main features: buildings, cross-streets, lampposts, trash cans, bike racks, benches, shop signs, trees, etc. (3) Try to recall details: numbers of windows, specific colors, any public art or posters, parked cars, unique doorways, even cracks in the pavement or plants growing in odd places. (4) Note any "landmarks" you always notice and anything you assume is always "the same." *Purpose:* This step recruits memory and the models your brain has built over repeated exposure—your implicit "map" of the environment.

Step 3: Note Your Confidence and Habits. Write down: (1) How confident are you in your recall of the details? (2) Which parts can you picture vividly and which are blurry or generic? (3) Are there parts you always rush past? (4) Are there spots you never really look at, only sense peripherally?

Step 4: The Ground Truthing Walk. (1) Take your sketch, map, or list and go to the street in person. Walk slowly, as if exploring this place for the first time. (2) With sketch in hand, compare what you thought was there with what you actually see. (3) What's different? What's missing or out of place? What details surprise you? (4) Are there features you've never noticed—graffiti, flyers, chipped paint, the shape of a sign, texture of the sidewalk, the presence (or absence) of birds, flowers, or puddles?

Step 5: Mindful Noting and Discovery. Pause frequently and note (aloud, in a notebook, or with your phone): (1) What assumptions did your memory "fill in" that aren't true? (2) Which details were left out? Which were invented or generic? (3) Are there changes you had never consciously registered (a new shop, a missing tree, a bench you never noticed)? (4) How does noticing these things feel—strange, uncomfortable, delightful, overwhelming? *Purpose:* This step helps reveal how automatic prediction (and inattentional blindness) shapes your lived reality.

Step 6: Reflect and Reintegrate. When you return home, reflect: (1) How different was your "mental map" from reality? (2) What does this reveal about everyday perception and memory? (3) How might you apply this insight to other parts of your life—relationships, work contexts, assumptions about people, etc.?

Step 7: Habitual Expansion. Set a micro-intention: "When I next walk this street, I'll look for three new details. Or, once a week, I'll pick a different familiar place to remap." *Purpose:* Over time, this cultivates not only mindful presence, but appreciation for novelty, reduces complacency, and boosts observational skills—practicing seeing "what is," not just "what I expect."

Optional Extension: Have several people (family, friends, colleagues) draw their version of the street, then walk together and compare perceptions. This can be quite fun and a wonderful social learning experience. Discuss how each person's "brain prediction machine" covers and misses different parts of reality.

The Street I Thought I Knew (A Story)

I've walked down Lincoln Avenue every morning for the last three years, always lost in thought about work, errands, or what to cook for dinner. I thought I knew the street as well as the back of my hand. So when I read about an exercise inviting me to draw my route before revisiting it, "just to see what you really notice," I was skeptical—and maybe a little smug. But I was curious.

[*Step 1: Drawing from Memory*] I sat down at the kitchen table, closed my eyes, and pictured the walk from my apartment to my favorite coffee shop on the corner. Storefronts blurred in my memory, more like signposts than real places: the bakery on the left, the laundromat with the green awning, the row of trees between 3rd and 4th. I drew it out, marking windows, doors, a couple of streetlights, and the "comic book shop" I'd never once entered. It was easy, almost automatic.

[*Step 2: Self-Doubt and Curiosity*] When I looked over my little map, I felt oddly uneasy. There were whole stretches I'd left blank. I couldn't remember if the postbox was before or after the bakery. Hadn't there been a flower stand once near the bus stop? I wrote a note: "Funny how 'familiar' feels empty when I really try to picture it."

[*Step 3: Walking with New Eyes]* Later that morning, I took my drawing and set off down Lincoln Ave. It was a cool, bright day, and people were out—neighbors, dog-walkers, a little girl trundling along behind her dad on a scooter. This time, I slowed down, paying attention.

Within half a block, I was surprised. The fence beside the laundromat wasn't plain gray as I'd drawn, but painted with a half-finished mural: bright blues and oranges, a cartoon sun peeking out of curlicued clouds. How had I missed that? Passing the bakery, I saw not the sign but the cracked tile doorstep, the tiny purple flower sprouting determinedly between bricks. The postbox was actually in front of the dry cleaner's, not the bakery. The comic book shop's windows were filled with not just comics but vintage toys and an artfully arranged stack of board games.

I saw flyers on the lamp post for a missing cat and an after-school yoga class. There was a tiny, cheerful wind chime on the door of the tailor's shop, chiming softly in the wind. Even the way sunlight moved on the shop windows seemed new.

[*Step 4: Realization]* At the coffee shop, I settled in and compared my drawing to reality. The differences were dramatic. My "map" was little more than a series of vague icons, shaped by habit and assumption. The real street was alive with details I'd never noticed before.

I realized then how much of my daily world my mind simply edited out. I'd assumed familiarity, but it was only expectation—my brain's "best guess" overlaying the rich, changing canvas of the real. What I called memory was more prediction than perception; what I labeled "seeing" was really a kind of strategic blindness that kept things efficient as I walked from here to there, but also, in some ways, shallow.

[*Step 5: Raised Awareness]* That evening, I noticed I was less "on autopilot." I found myself pausing at the new mural, actually looking at the faces of people who passed, feeling a flicker of curiosity—where were they headed? What stories filled their day? The world felt sharper, as if someone had fiddled with a knob and turned up the color and clarity.

I understood now that "expanding awareness" isn't about grand revelations—just opening, moment by moment, to what I usually miss. It's a practice. One I'd nearly missed, right in front of me.

Takeaway: Through a simple act of mapping from memory and then really looking, the street became new again. What started as a challenge to my memory led to a challenge of my awareness itself. I began to appreciate the subtle richness of the present, and how much possibility hides in plain sight.

After Thoughts

Let's quickly synthesize our learning from Hawkins' Thousand Brains theory. The neocortex is composed of many similar modules, each building models of the world—by mapping and predicting structure and relationships (reference frames). We perceive what is "right in front of us" only when the collective output of these modules is integrated and matches our attention and goals. The intelligent brain is always testing predictions against reality, and can miss obvious details if they don't "fit" its mental models. Thus, "blindness" to the obvious may occur when these models align on an incorrect prediction, leading incoming information to be ignored, misinterpreted, or suppressed.

And two more points to remember. First, awareness is not passive "seeing," but an active process of inference, selection, and integration. Second, most data never reaches conscious perception because it is filtered by attention, expectation, or deemed irrelevant to current goals.

In the next chapter, we take an interdisciplinary deep dive into attention. Such an approach does more than aggregate knowledge; it creates new insights at the intersections, focusing on facets of the question that would potentially remain obscure within any single perspective. By weaving together understandings of such things as attention, beliefs, cognitive dissonance, perceptual blindness, biases, and self-deception as understood *across* disciplines, we move closer to a multidimensional answer. In doing so, we not only deepen our grasp of why seeing (and not seeing) happens as it does, but we also empower ourselves to cultivate a broader, more flexible lens for understanding the human experience.

And so—with the foundation set for our discovery—we begin a deeper dive to explore the question: *Why can't we see what is right in front of us?*

Chapter 7
Attention and Intention

In the dance where attention spans play,
Focus frequently does go astray.
With intentions aligned,
We *Whysly* will find
New paths where distractions give way.

Attention—a tool of the self directly related to consciousness—is essential for bringing stimuli into conscious awareness. It lays the web for interaction with the world in which we act. The emerging interdisciplinary field of attention sciences combines insights from psychology, neuroscience, and cognitive science to study attention's mechanisms and effects. Researchers explore various forms of attention such as divided, sustained and selective attention, providing a comprehensive understanding of how and why we focus on specific stimuli.

From a neuroscience perspective, attention emerges as the dynamic orchestration of various brain regions working together to filter, prioritize, and select information for conscious awareness. The *frontal cortex*, particularly the prefrontal areas, acts as an executive hub, setting goals, regulating focus, and allocating attentional resources based on task demands and relevance. Simultaneously, the *parietal lobe* plays a crucial role in spatial awareness—helping us zero in on specific locations in our environment, almost like an internal spotlight. The *anterior cingulate cortex* is pivotal in monitoring for conflicts (such as distractions or competing stimuli) and helps mediate the switch between different attentional states. Functional MRI studies demonstrate that these regions become especially active during tasks that require sustained focus or when we must switch attention rapidly, as in busy, information-rich settings. Essentially, these networks govern what is granted access to conscious perception and what remains outside our awareness (Gazzaniga, 2018).

Crucially, *intention*—our goals and expectations—directly shapes how these brain systems allocate *attention*. When we set an intention, say to find a friend in a crowd, our prefrontal cortex communicates this priority to other regions, biasing both what we notice and how we interpret sensory input. This process is often referred to as *top-down modulation*: what we hold in mind actively influences what our brain treats as "important." Neural pathways facilitate this bias, making us more likely to perceive elements that match our current intentions or frameworks, and less likely to register unexpected or irrelevant details. As a consequence, even if something is right before our eyes, it may go unnoticed if it does not fit the narrative or expectation our brain is primed for. This selective processing, while efficient, inevitably creates "blind spots" in perception—explaining why our subjective reality is so often shaped by what we expect or hope to see, rather than by the totality of what is actually present.

Attention can only occur in the NOW, and sustained attention is a series of NOWs. For example, the eyes work much like an old-time movie reel, with a fixation occurring approximately three times a second as the eyes make a small, quick movement (a saccade) and then stop. However, we perceive it as a continuous flow of movement. Just as the eyes shift their focus from object to object, attention shifts as driven by our thoughts and feelings. When these thoughts and feelings are focused and consistent, we are effectively creating the reality of the moment in which we exist (Bennet, Bennet, & Turner, 2022).

Although our eyes and minds are constantly gathering information, we don't process every detail available to us. Instead, attention acts as a gatekeeper, filtering the overwhelming flood of sensory data into manageable, meaningful moments—the "NOWs" we experience. This process, like the quick saccades of our eyes, is both rapid and selective. As our focus shifts, guided by internal thoughts and emotions, certain things fall outside the spotlight of attention. This selective focus explains why we often fail to notice what is right in front of us—creating these so-called "blind spots" in our perception. In essence, what we don't attend to, we don't consciously see, even if our eyes are looking right at it.

Divided, Sustained and Selective Attention

Understanding divided, sustained, and selective attention is essential for grasping how cognitive limitations can lead to failures in recognizing significant changes or details in our perception. Each type of attention plays a distinct role in shaping our focus and awareness, highlighting why we may not "see" what is right in front of us.

Divided attention refers to the ability to distribute cognitive resources across multiple tasks or stimuli simultaneously. This form of attention is critical in situations where individuals must multitask, such as driving while talking on the phone. However, research indicates that dividing attention among tasks can lead to a decrease in performance and the likelihood of missing significant information (Salvucci & Taatgen, 2008).

Divided attention is crucial for understanding "not seeing" because when cognitive resources are spread too thin, important changes in the environment may go unnoticed. For instance, a driver conversing on a phone may fail to see a pedestrian crossing the street, indicating how limitations in divided attention can result in oversights or "blind spots".

Scenario Example: Imagine a busy parent preparing dinner while helping a child with homework and occasionally glancing at their phone. The cognitive demands of cooking, supervising the child's assignments, and checking messages constantly pull their attention in different directions. In the midst of juggling these tasks, the parent doesn't notice the pot on the stove boiling over, even though it's right in their direct line of sight. In this scenario, the act of dividing attention among multiple concurrent tasks leads to an oversight. The visible signs of the boiling pot are ignored because cognitive resources are stretched too thin, exemplifying how dividing attention can result in missing crucial information that's literally right in front of us.

Sustained attention, also known as vigilance, refers to the ability to maintain focus on a task or stimulus over an extended period. This form of attention is essential in contexts such as monitoring situations for changes (e.g., watching a surveillance screen or focusing on a lecture) or completing tasks that require a prolonged cognitive effort (Parasuraman & Hsu, 2000).

Sustained attention is significant in understanding why we might miss changes in our environment. If an individual's sustained attention wanes due to fatigue or monotony, they may fail to recognize minor but significant changes occurring over time. A classic example is how an observer may not notice a light slowly dimming while engaged in another task.

Scenario Example: Picture a security guard assigned to monitor several surveillance screens in a quiet building during a long, slow night shift. For hours, nothing of interest happens, and the monotony gradually erodes the guard's alertness. A subtle change—like an unauthorized person briefly appearing in a dim corner of one camera feed—goes unnoticed. In this case, the guard's prolonged effort to maintain vigilance becomes compromised over time, and their sustained attention falters. Even though the incident happens right before their eyes, the failure to maintain attention throughout the duration results in missing the event.

Selective attention is the cognitive process of focusing on specific stimuli while ignoring others. It allows individuals to prioritize information that is deemed relevant while filtering out distractions. This mechanism is crucial for reducing cognitive overload in environments with multiple competing stimuli (Treisman, 2006).

Selective attention is vital for understanding why we might not see what is right in front of us because it dictates where our focus lies. When attention is directed selectively, individuals may fail to notice important details that do not align with their current focus or expectations. For instance, during a complex conversation, one might overlook an important visual cue in their peripheral vision because their attention is fixed on the speaker.

Cognitive psychology offers a foundational explanation for selective attention through Broadbent's "filter theory." According to this model, our cognitive system acts like a gatekeeper. Early in the processing stream, it filters incoming sensory information, allowing only the most relevant or attended inputs to proceed to conscious awareness, while the rest is effectively blocked or discarded. This means that even though we are constantly bombarded by a wealth of sensory details, only what we focus on enters our conscious perception; information residing in unattended channels is simply not processed at a level sufficient for awareness. Broadbent's theory directly explains why we can fail to see what is right in front of us: changes or important details that occur outside our locus of attention may be missed entirely, not because our eyes failed to register them, but because our minds did not let them through for conscious recognition (Broadbent, 1958).

Cultural studies deepen our understanding of selective attention by highlighting how cultural contexts shape both what we notice and what we ignore. Cultural norms, values, and shared narratives inform our intentions and subtly guide our attentional priorities. For instance, in collectivist cultures, individuals often attend more closely to social cues and group dynamics, sometimes at the expense of noticing individual details. In contrast, cultures that value individualism may foster a focus on personal achievements or traits, potentially overlooking broader social contexts. These culturally ingrained patterns create selective "attentional filters," guiding what is perceived as significant while relegating other stimuli to the background. As Stuart Hall (1997) points out, cultural representations influence both our intentions and expectations, which in turn determine what draws our attention. Thus, cultural context can lead us to regularly overlook details and changes that fall outside our society's dominant narratives—meaning that even when something is directly in front of us, it may remain unnoticed if it is not culturally marked as meaningful.

Scenario Example: Consider an office worker engrossed in drafting an urgent report, focusing intently on the words glowing on their computer screen. While typing, a colleague enters the room and subtly sets an important document on the worker's desk. Because the worker's attention is selectively allocated to the report and the task at hand, they remain oblivious to the new document's presence, even though it's placed directly within their view. Here, selective attention guides perception so narrowly that obvious elements in the environment—those not deemed "relevant" to the immediate task—can be entirely overlooked

The Attentional Blink

The attentional blink is a phenomenon in cognitive psychology that refers to a brief period after the perception of a stimulus during which a second stimulus cannot be perceived, even if it is presented visually. This effect typically occurs when two target stimuli are shown in rapid succession within a visual display. Research indicates that if the second target appears within approximately 200–500 milliseconds after the first, observers often fail to notice it—a finding that illustrates the limitations of attentional capacity (Rayner & Castelhano, 2008).

The concept of the attentional blink provides a vivid example of how attention operates in discrete "moments" or "NOWs," rather than as a seamless, continuous process. Just as our eyes fixate briefly and then move, creating the illusion of uninterrupted vision, our attention can only process so much at once. The attentional blink demonstrates that after we process one stimulus, there is a short window—a kind of cognitive refractory period—during which our attentional resources are occupied and a second stimulus may be missed entirely. This reinforces the idea that at any given moment, we are selectively constructing our reality based on what captures our attention, and that our perception of a continuous flow is built from these separate, rapidly shifting points of focus. In essence, the attentional blink lays bare the limitations of our "series of NOWs," reinforcing that our awareness truly depends on both what we attend to and when we attend to it.

The attentional blink is critical for understanding the challenges related to "not seeing" important information because it highlights the temporal limitations of attention. During this narrow window post-initial perception, cognitive resources are reallocated to process the first stimulus, effectively reducing awareness of any subsequent stimuli. This phenomenon has real-world implications; for example, when multitasking or engaging in rapid visual scanning, individuals may overlook critical changes or cues that follow immediately after recognized stimuli.

Scenario Example: Imagine a radiologist rapidly interpreting a series of medical images, such as X-rays, searching for two types of abnormalities: a tumor (target one) and a subtle fracture (target two) that sometimes appear close together in quick succession. Upon detecting the suspicious mass, the radiologist's attention is fully engaged with processing and evaluating it, and in those critical next few hundred milliseconds, their capacity to perceive the second abnormality is diminished. As a result, the faint fracture—though present and visible on the screen—may go unnoticed entirely. This scenario showcases the attentional blink in action: the very act of perceiving and responding to one important stimulus temporarily renders a professional blind to another, equally critical piece of information that follows closely behind.

The attentional blink illustrates the constraints of cognitive processing and highlights a specific scenario in which individuals may fail to notice vital information right in front of them. By examining this phenomenon, we deepen our understanding of how attention operates in time-sensitive contexts, emphasizing the interplay between attention, perception, and cognitive limitations.

Attention Restoration Theory

Attention Restoration Theory posits that natural environments can help restore depleted attentional resources through exposure to nature, allowing individuals to recover from built-up attentional fatigue. The theory highlights the beneficial effects of restorative experiences on cognitive functioning. For example, after spending a long day in a hectic office, taking a walk in a serene park filled with trees and flowers can help rejuvenate your focus and make you more aware of your surroundings upon returning to work.

This theory is important in understanding attention, as it suggests that prolonged focus in artificial environments (like urban settings) can lead to cognitive fatigue, which may heighten the risk of missing important stimuli. Engaging with nature allows individuals to refresh their attentional capacities, enhancing their ability to notice changes upon returning to more demanding environments.

Kaplan (1995) specifically focuses on the concept of restorative environments and their impact on attentional capacity, providing evidence that natural settings can enhance cognitive functioning and improve attentional restoration after fatigue. This framework underscores how environmental contexts influence attention, supporting that exposure to nature can help.

This theory is crucial to understanding why we often fail to notice things that are right in front of us because it highlights the effects of cognitive fatigue on our perceptual abilities. In our increasingly distracting and fast-paced environments, such as urban settings, our attentional resources become depleted. This fatigue can impair our ability to process and integrate information effectively, leading to oversights of important details in our surroundings.

Scenario Example: Imagine a busy city street where the constant hustle and bustle overwhelms your senses, demanding your full attention to navigate through throngs of people and blaring traffic. As you rush through this urban landscape, your mind becomes fatigued, sapped of the ability to process information effectively. Now picture this: after a long day of managing distractions and focusing on tasks, you decide to step into a nearby park. The lush greenery, gentle sounds of rustling leaves, and the vibrant colors of flowers create a restorative environment that allows your mind to unwind. This exposure to nature rejuvenates your attentional resources, enabling you to refresh your cognitive capacities. When you return to the city, you're surprised to suddenly notice details you overlooked before—the unique architecture of buildings, the subtle expressions of passersby, or even the scent of fresh coffee wafting from a nearby café. This transformation speaks to the essence of why we often fail to see what is right in front of us: prolonged attentional fatigue in overstimulating environments can dull our perception, while restorative experiences in nature reawaken our awareness, allowing us to appreciate the richness of our surroundings.

Ecological Psychology

Ecological psychology emphasizes the relationship between individuals and their environments, highlighting how environmental aspects affect perception and attention. This field posits that organisms perceive their environment through affordances—opportunities for action which are directly perceived. The ecological perspective suggests that attention is influenced by the environment and that changes in context can provide essential cues to what individuals attend to. If the environment does not afford noticing a particular change, it can lead to failures in perception and recognition.

Gibson (1979) discusses just how perception is grounded in the interaction between organisms and their environments. He emphasizes that attention is shaped by ecological contexts, influencing what is perceived as relevant or significant. This underscores the idea that if an environment does not effectively signal changes, individuals may fail to notice important details in their surroundings.

Ecological psychology offers valuable insights into why we often fail to notice what is right in front of us by emphasizing the interplay between our perception and our environment. According to this perspective, we perceive objects and events not merely as isolated entities but through the lens of affordances—that is, the opportunities for action they present. If an environment does not signal or afford certain cues, critical changes or details can go unnoticed. For example, if the layout of a room remains static and does not highlight a new piece of art or furniture, individuals may overlook these changes entirely. This means that our attentional focus is not solely a matter of cognitive processing; it is profoundly influenced by how our surroundings frame what we perceive as relevant. Consequently, if our environment lacks the appropriate signals to direct attention, we are likely to miss important information right before us.

Scenario Example: Imagine a bustling coffee shop where patrons are immersed in their laptops and conversations. A new, eye-catching mural is painted on the wall, but its introduction coincides with a particularly busy time of day. Customers are preoccupied with phone calls, laptop notifications, and the barista's orders. As they sip their coffee, none of them engage with the mural, as their environment is not conducive to noticing such a change. The hustle and bustle acts as a backdrop that does not afford attention to the mural, leading to its existence being completely overlooked. In this scenario, the failure to notice the mural illustrates the core idea of ecological psychology: our perception is deeply tied to environmental contexts, and when these contexts do not signal changes effectively, we may fail to recognize significant elements right in front of us.

Explicit and Implicit Attention

Epistemology, the branch of philosophy concerned with the nature of knowledge, belief, and justification, touches significantly upon the concepts of explicit and implicit attention. Explicit attention refers to the conscious, deliberate focus on specific stimuli or information, while implicit attention operates outside of conscious awareness, influencing how we recognize and prioritize information without deliberate effort. These concepts underscore how we process knowledge and shape our understanding of our environment (Atran & Medlin, 2008).

The distinction between explicit and implicit (tacit) attention is crucial for understanding why we might not see what is directly in front of us. When information is categorized as explicit, it is typically the focus of our conscious awareness. However, implicit or tacit (as it is called in the field of Knowledge

Management) attention affects our interactions with the environment in ways we may not fully recognize or control. For example, if an observer is consciously focused on reading a text (explicit attention), their implicit attention might still process peripheral cues, but significant changes in their environment may go unnoticed if those changes do not align with their current focus. This can explain why people often "fail to see" important details—when they may not be explicitly attending to them.

Scenario Example: Imagine a museum visitor intently reading the detailed description beside a famous painting. Their explicit attention is fully engaged with the text, as they strive to understand the historical and artistic context. Meanwhile, their implicit attention operates in the background, vaguely registering the movements of other visitors and the ambient lighting in the room, but not consciously noting the gradual unveiling of a new artwork in a corner display. Despite the new piece being in plain sight, the visitor doesn't notice it until someone else draws their attention to it. This scenario illustrates how explicit and implicit attention work together—explicitly, our focus is guided by immediate goals or interests, while implicitly, we may absorb peripheral information without conscious awareness. However, when something falls outside the scope of our explicit focus and fails to trigger implicit relevance, it can remain unseen, demonstrating how these attentional processes shape and limit our experiential reality.

The explicit and implicit connection to attention within epistemology reveals important mechanisms that influence perception and awareness. By understanding how conscious and unconscious processes operate in tandem, we can better grasp why individuals may overlook details or changes that seem apparent, highlighting significant philosophical insights into the nature of knowledge and perception.

Feature Integration Theory

Feature Integration Theory (FIT), developed by Anne Treisman (1986), explains how individual features of objects (such as color, shape, and size) are processed separately before being integrated into a single perception. The theory posits that focused attention is necessary to bind these features together into a coherent perception.

FIT shows that without focused attention, distinguishing between objects and their features can fail, leading to oversights. If attention is diverted, critical components that inform recognition may not be integrated, resulting in unrecognized changes or details—even if they are physically present.

Treisman's research investigates how attention is necessary for integrating features into a single perceptual object. She explains that when focus is divided or lacking, individuals may fail to correctly perceive salient features, leading to important details going unnoticed. This work underlines the role of attention in acknowledging changes in our environment, reinforcing the implications for "not seeing" what is right in front of us.

Scenario Example: Imagine you are walking through a bustling market filled with vibrant stalls. As you pass by a fruit vendor, you momentarily become distracted by the lively conversation of your friends, who are animatedly discussing an upcoming trip. While your attention is diverted, you glance at a table filled with various fruits. You notice a bright red apple and a large yellow banana, but as you continue talking, you fail to fully fixate on these items.

Later, you return to the same stall with your friends and are surprised to see that, in addition to the apple and banana you initially noticed, the vendor has displayed a striking arrangement of exotic fruits, including a dragon fruit and a starfruit. Because your attention was divided during your first glance, you had not integrated all the available features of the scene into a coherent perception. This oversight illustrates how the absence of focused attention can lead to missing important details—like the new fruits—despite them being physically present in front of you. The experience underscores Treisman's Feature Integration Theory, highlighting the essential role of attention in enabling us to perceive and recognize the full richness of our environment.

Psychobiology of Attention

Psychobiology examines the biological underpinnings of psychological processes, including attention. This field studies how physiological responses and neurochemical processes impact attentional mechanisms and cognitive functioning. Understanding the psychobiological aspects of attention reveals how physiological states—such as stress or fatigue—can influence attentional capacity. For example, heightened stress can lead to a narrower focus, making individuals less likely to recognize changes in their periphery or surroundings due to a constrained attentional span.

Raine and Chen (2017) provide insights into how biological factors influence attention and behavior. By exploring the neurobiological underpinnings of attentional processes, their work illustrates how physiological conditions can impact cognitive resources, indicating that impairments in attention due to biological factors can lead to missed information.

The psychobiology of attention is crucial in understanding why we often miss important stimuli right in front of us, as it highlights how our physiological states influence our cognitive functioning. Factors such as stress, fatigue, and emotional states can significantly impact how we allocate our attentional resources. For instance, when we are under heightened stress, our focus tends to narrow, which can severely limit our ability to perceive information outside of our immediate focus. This physiological response leads us to inadvertently overlook changes or important details in our environment, as our cognitive capacity becomes constrained. Thus, when our attentional mechanisms are influenced by biological factors, we risk being blind to what is happening in our surroundings, reinforcing the idea that stress and other physiological states can create blind spots in our perception.

Consider a scenario in a busy hospital emergency room. A nurse is overwhelmed by a sudden influx of patients, leaving them under significant stress. As they dart between rooms, focused on triaging the most critical cases, they are fixated on the immediate needs in front of them. A doctor standing just a few feet away raises a hand to signal an urgent matter, but the nurse, overwhelmed and stressed, fails to notice this important cue. Their narrowed focus, driven by the high-pressure environment, leads them to miss a key piece of information that could affect patient care. In this case, the physiological stress response has directly impacted their attentional capacity, illustrating how biological factors can result in missing crucial stimuli that are literally right in front of them.

Ritualized Attention

Ritualized attention refers to the focused awareness and deliberate practices involved in performing rituals, often characterized by repetition, symbolism, and heightened sensory engagement. For example, a monk engaging in a meticulous tea-making ceremony that requires precise movements and mindful presence.

In his research, Roy Rappaport (1999) explores how rituals function as pivotal components in human societies, transcending simple routines to facilitate the construction of social truths and shared realities. Rappaport argues that rituals serve as mechanisms for maintaining social order, transmitting cultural values, and coordinating group activities. He emphasizes that ritualized attention—the sustained focus and commitment during rituals—is vital for imbuing these practices with meaning, reinforcing community bonds, and stabilizing cultural narratives. By centering attention, rituals transform ordinary actions into symbolic acts that convey deeper cultural and social significance.

Ritualized attention contributes to understanding why we might overlook what is immediately before us by demonstrating how cognitive and perceptual priorities are shaped through cultural practices. Rituals often require participants to adopt a specific focus, directing attention towards symbolic acts and shared meanings. This deliberate attention can foreground particular elements while backgrounding others, revealing how cultural practices influence what we perceive as significant. When people are deeply engaged in ritualistic activities, their attention is trained on prescribed elements, which might lead them to miss spontaneous or unscripted occurrences outside the ritual's framework. This suggests that culturally ingrained habits of attention can affect perceptual awareness in ways that obscure certain realities.

Scenario Example: During a traditional wedding ceremony in a small village, the bride and groom sit beneath a beautifully adorned canopy, surrounded by family and friends, who are all dressed in vibrant attire. As the officiant begins a series of prayers and blessings, the attendees focus intently on the couple, their words, and the various symbolic rituals being performed, such as the lighting of a unity candle and the exchange of garlands. Everyone is deeply engaged in this ritualized moment, their attention drawn to the sacred vows being exchanged and the shared cultural significance of the ceremony. However, amid this concentrated focus, the spontaneous laughter of children playing nearby or the soft chatter of guests discussing their favorite dishes at the reception goes unnoticed, illustrating how the ritualized attention on the wedding transforms the event into a meaningful celebration while obscuring the lively, unscripted moments occurring just outside the circle of engagement.

Social Attention

Social attention refers to the way our focus is influenced by social contexts, including the presence, behaviors, and expectations of others around us. Unlike purely individual forms of attention, social attention is shaped by group dynamics, social cues, and cultural norms. For instance, people are naturally drawn to look where others are looking, follow the gaze or gestures of peers, or pay more attention to people they feel connected to. This powerful pull can cause our priorities and perceptions to shift in subtle but important ways depending on who is around and what social signals they send.

Within the context of why we often fail to see what is right in front of us, social attention reveals that our perceptual blind spots are not simply the result of individual cognitive limitations, but are also shaped by group affiliation and social signaling. For example, we may unconsciously tune out information from people we perceive as outsiders (out-group members) or over-prioritize signals from those with whom we closely identify. This can lead to both oversight and bias—critical details may be missed or ignored simply because they originate from someone outside our social circle, or they do not align with the collective focus of the group.

Scenario Example: You attend a team meeting at work where most members share similar backgrounds and have known each other for years. A new colleague, unfamiliar to the group, raises a concern about a possible error in the project plan. Despite speaking up, the new member's input goes largely unnoticed; the rest of the team remains focused on the familiar leader's presentation and ongoing discussion. Here, social attention is at play: the group's shared focus and implicit social biases cause them to overlook a potentially critical piece of information that is, quite literally, being presented in front of them. This scenario illustrates how social context can effectively shape, and sometimes limit, what we perceive and attend to.

The Bystander Effect. A related phenomenon in social psychology is the bystander effect, which describes how the presence of others can dilute an individual's sense of responsibility and focus, especially in situations that may require action or vigilance. When multiple people witness an emergency or unusual event, individuals are less likely to take notice or intervene, often assuming that someone else will respond. This diffusion of attention and responsibility powerfully demonstrates a social mechanism underlying failures to perceive and react to what is plainly in sight.

The bystander effect is deeply relevant to our theme because it demonstrates that perceptual and attentional blind spots are not just internal or private phenomena, but are also constructed by the social environment. In situations where multiple individuals are present, the responsibility—and therefore the attention—each person feels is diluted. As a result, our attention can become diffused or even paralyzed, making us less likely to recognize or respond to urgent situations, even if they occur in plain sight. This social dynamic illustrates how, in the presence of others, people can collectively fail to see or acknowledge what is happening right in front of them; our tendency to look to others for cues can override our own perception, contributing to inaction or oversight in critical moments.

Scenario Example: Consider a crowded subway platform where someone collapses and lies motionless. Dozens of commuters witness the event, but each hesitates, glancing around to gauge others' reactions. Because no one immediately responds, many fail to register the urgency or seriousness of the situation, rationalizing that someone more qualified or responsible will help. In this scenario, the presence of the group fragments attention and intention, causing everyone to overlook the immediate need—which is right before their eyes. The bystander effect, therefore, starkly illustrates how our social environment can create collective blind spots, leading us to miss or ignore critical details or events even in the midst of a crowd.

Temporal Attention

Temporal attention relates to how individuals allocate their attentional resources over time, focusing on specific moments when they expect important events to occur. Research has shown that temporal attention affects the processing speed and accuracy of stimuli encountered at different times. If attention is temporally aligned with specific expectations, other events occurring during non-focused intervals may be overlooked. For example, if one is attuned to an expected sound (like a phone ring), they may not notice a visually significant event happening simultaneously.

Coull and Nobrer (2008) investigate how temporal and spatial attention can be disentangled, focusing on how attentional resources are allocated across time. The results indicate that temporal attention can significantly enhance processing for expected stimuli while compromising awareness of concurrent visual changes. This framework highlights how attending to time-focused cues can lead to not seeing other critical stimuli present in the environment.

Temporal attention plays a significant role in understanding why we often fail to notice important stimuli right in front of us. When we align our attention with specific moments based on our expectations—such as waiting for a phone to ring or anticipating an important announcement—we unconsciously allocate our cognitive resources to those anticipated events. This laser-like focus means that we may become oblivious to other significant stimuli occurring simultaneously. For instance, while our ears might be attuned to the sound of a ringing phone, our eyes could completely miss a colleague's subtle yet important gesture or a change in the environment, like the arrival of someone we should greet. Consequently, the very act of prioritizing our attention temporally can lead to blind spots in our awareness, making it easy to overlook what is right in front of us.

Scenario Example: Consider a scenario in a busy office during a critical meeting. Everyone is seated around a conference table, and the team leader is about to unveil a crucial presentation. As the presenter prepares to speak, a participant checks their phone in anticipation of an important text message related to the project being discussed. Their conscious attention is solely focused on the screen as they await the alert, leading them to miss essential visual cues in the room: a colleague raising their hand to contribute an insightful idea, or the projector displaying a graph that could significantly impact the discussion. This moment underscores the essence of temporal attention—by prioritizing their attention on a single expected event (the phone alert), the participant inadvertently becomes blind to other critical interactions occurring right in front of them.

Contributions from Other Fields and Studies

Each of the fields and studies called out below bring a unique perspective on why we "miss" what' right in front of us—whether due to design choices, psychological states, environmental factors, technology interfaces, or even philosophical differences. We'll provide a scenario example for each.

Learning sciences and educational psychology examine the ways that attention shapes the process of acquiring and retaining knowledge. Researchers study how factors such as student engagement, motivation, environmental distractions, and instructional design impact learners' ability to focus, process, and internalize information. Attention is recognized as a finite resource, and educators often employ strategies to optimize students' attention spans and reduce distractions in the classroom (Moray, 1969).

One of the main challenges in education is that students often fail to absorb important information—even when it's presented clearly—simply because their attention is pulled elsewhere. For example, a student concentrating on taking notes may miss the teacher's demonstration on the board, not due to a lack of capability, but because their attentional resources are already consumed. Thus, the design of learning environments, teaching methods, and even classroom layouts all play crucial roles in determining whether students "see" and remember what's directly in front of them.

Scenario Example: A high school student sits at their desk, focused intently on copying vocabulary words into a notebook. Meanwhile, the teacher demonstrates a critical math concept at the front of the class. Despite being physically present and looking toward the board, the student misses the key visual demonstration, later struggling on homework problems related to it. Their attention, directed by the immediate task of note-taking, prevented them from registering the live instruction just a few feet away.

Human-Computer Interaction (HCI)/Design. HCI investigates how users interact with computers, digital devices, and interfaces, with particular attention to how design elements guide, capture, or fragment user focus. The interface layout, color schemes, notifications, and even animation are crafted to draw attention toward certain features and minimize cognitive overload. Increasingly, the "attention economy" raises concerns about digital distractions and interruptions (Norman, 2013).

Our digital environments are filled with competing stimuli, each designed to draw our focus. As a result, users can easily miss crucial information on a screen—such as a warning message or important button—because their attention is captured by more prominent or visually stimulating elements. Design missteps, such as poor placement of information or excessive notifications, directly contribute to why users may "not see" something present and essential in the digital interface.

Scenario Example: A user receives a security alert on their online banking website, but the notification appears as a small, grey banner above a brightly colored promotional ad. Their attention is drawn to the flashing ad at the center of the page, and they fail to notice the critical alert at the top—leaving their account momentarily vulnerable.

Clinical Psychology/Psychiatry. Clinical psychology and psychiatry focus on how mental health and neurological disorders influence attentional capacity, control, and stability. Conditions such as ADHD, anxiety, trauma, and depression affect attentional processes, often resulting in hyperfocus or distractibility. Therapies—such as cognitive-behavioral therapy or mindfulness—may be employed to help individuals retrain or better manage their attention (Barkley, 2015).

Individuals with attention disorders may frequently miss obvious cues in their environment; a person with ADHD, for instance, may not register a friend waving at them across a room, simply because their attention is momentarily absorbed by their thoughts or other stimuli. Mental health challenges can make it even more difficult for people to "see" what's right in front of them, due to internally driven attentional biases or disruptions.

Scenario Example: A young adult with generalized anxiety disorder is preoccupied with internal worries during a team meeting at work. While physically present, they do not notice when their supervisor asks them a direct question until a colleague points it out, having been absorbed by their anxious thoughts rather than external events—demonstrating how psychological state can override awareness of immediate surroundings.

Marketing and Advertising. Marketing and advertising are fundamentally invested in capturing and directing consumer attention. This field leverages color psychology, content placement, repetition, and imagery to win moments of consumer focus in increasingly crowded media landscapes. Techniques such as product placement and "call-to-action" elements are designed specifically to stand out amidst noise and competition (Kahneman, 2011).

Because marketers are constantly vying for attention, they also reveal how easy it is for important information—such as the terms and conditions of a product, or side effects of a medication—to be overlooked. Consumers often miss significant warnings, disclaimers, or product details right in front of them, as their attention is steered by enticing visuals or slogans intentionally designed to overshadow less marketable but crucial information.

Scenario Example: A viewer watching a pharmaceutical ad is captivated by bright visuals and uplifting music showcasing happy people. Meanwhile, critical information about possible side effects scrolls by in small print at the bottom of the screen. The engaging imagery means the viewer "misses" vital safety information placed right in front of them.

Ergonomics/Human Factors Engineering. Ergonomics (or human factors engineering) designs tasks, equipment, and workspaces around human cognitive and physical capabilities, including attentional limits. In safety-critical fields such as aviation, medicine, or driving, attention is a finite and precious resource. Human-factors experts analyze how to reduce errors, design warning systems, and arrange controls and displays for optimal attention allocation (Wickens et al., 2015).

Accidents and oversights in high-stakes environments often occur not because warning signs or hazards are invisible, but because human operators do not notice or process them—typically due to attentional overload, poor interface design, or multitasking. The field's focus on minimizing "attention capture failures" underscores how the right information in the wrong place can go unseen.

Scenario Example: An airline pilot is managing multiple alarms during stressful weather conditions. Despite a warning light appearing on the instrument panel, their attention is locked on the turbulence controls, so they only notice the alert when it begins to sound an audible alarm—demonstrating how even critical information, if not effectively highlighted or prioritized, can be overlooked.

Linguistics and Psycholinguistics. Linguistics and psycholinguistics study how we process language in real time, including how attention guides our understanding of speech, reading, and conversation. Researchers look at mechanisms like garden-path sentences, where initial interpretations are rapidly revised, and investigate how prosody or visual cues direct listener and reader attention to meaning (Altmann, 1997).

When reading or listening, our attention is constantly guided by context, grammatical structure, and expectation. This can lead to misinterpretations or "not seeing" (or hearing) what is present; for instance, skipping over ambiguities or ignoring corrections if they come from unexpected speakers. Linguistic cues that don't align with our expectations may simply be filtered out—and their importance missed.

Scenario Example: During a complex sentence, "The old man the boats," a reader quickly interprets it as describing an "old man" rather than parsing the ambiguous structure. The reader's attention, guided by common patterns, overlooks the sentence's actual meaning (that "old" is the subject and "man" the verb)—showing how attentional expectations shape what we notice in language.

Artificial Intgelligence and Robotics. In AI and robotics, scientists develop systems that must prioritize and process visual and sensory information much like humans do. The concept of "artificial attention" allows these systems to identify, select, and respond to relevant stimuli in environments that are often cluttered with competing data points. Attention models are integral to tasks like image recognition, navigation, and human-robot interaction (Vaswani et al., 2017).

AI's "blind spots," and those in computer vision, mirror human limitations: a robot or algorithm might fail to detect an object that is technically present in its input stream because its attention module is tuned toward more salient patterns or programmed priorities. This not only informs our understanding of machine intelligence, but also provides insights into the selective nature of human attention and perception's limitations.

Scenario Example: A self-driving car is equipped with sensors and cameras to interpret its environment. As it drives, its algorithms prioritize recognizing lane markings and moving vehicles but fail to register a small, stationary object (such as a fallen bicycle) on the road ahead—demonstrating that even with advanced technology, attention allocation determines what is truly "seen" and responded to.

Mindfulness and Contemplative Studies. Mindfulness and contemplative disciplines investigate practices designed to enhance conscious, sustained attention and awareness of the present moment—often developed through meditation, yoga, or contemplative prayer. Research in this field examines how intentional focus can be cultivated and how it supports well-being, emotional regulation, and cognitive flexibility (Kabat-Zinn, 2013).

One of the key promises of mindfulness is the reduction of "blind spots" in perception: by training the mind to return to the present and maintain open awareness, practitioners aim to perceive more of what is truly present in each moment, rather than being lost in distraction or automatic filtering. Nonetheless, mindfulness practice also reveals just how persistent and pervasive inattentional blindness and habitual distraction can be in everyday life.

Scenario Example: During a mindfulness meditation, a participant notices their mind repeatedly wandering to future plans, missing the subtle sounds, sensations, and sights of the present. With practice, they begin to register more of these immediate details, yet are surprised by how much they previously failed to notice—demonstrating firsthand how untrained attention can make us miss what's right in front of us.

Intention

Intention can be defined as a mental state that represents a commitment to carrying out an action or achieving a specific goal. It is a cognitive process that involves planning and motivation, where individuals set their sights on a desired outcome and actively or passively pursue it. Intentions guide behavior by influencing the choices people make, shaping their focus, and determining the resources they allocate to achieve their objectives. This mental framework is more than a passive wish for a future state; instead, it embodies an active element of purpose and determination, setting the stage for action.

The importance of intention in human life extends far beyond goal-setting. It plays a crucial role in personal development, decision-making, and social interactions. Intentions help individuals navigate complex environments by providing a sense of direction and clarity, allowing individuals to prioritize their actions in alignment with their values and aspirations. Furthermore, intentions foster a sense of agency; when people set intentions, they feel empowered to influence their outcomes, enhancing motivation and resilience in the face of challenges. Overall, intention serves as a cornerstone of human behavior, shaping how individuals perceive their world, interact with it, and ultimately create the lives they wish to lead.

These additional frameworks from various fields—including behavioral economics, education, mind-body medicine, human-computer interaction, and psycholinguistics—offer unique insights into the role of intention and its relationship to attention and perception. By exploring how intention shapes focus and awareness, we can better understand the complexities that contribute to the phenomenon of not seeing or recognizing important details in our environment.

Behavioral Economics. Behavioral economics studies the psychological, cognitive, emotional, and social factors that influence economic decisions. It explores how intentions and cognitive biases affect decision-making processes, often leading individuals to focus on certain outcomes while ignoring others (Thaler & Sunstein, 2008). Intention within this context can shape how individuals allocate their attention towards specific choices or preferences, often overriding less salient but important information. For example, a consumer intent on finding the best deal may wholly focus on price comparisons while overlooking quality indicators or potential risks, leading to missed important details that could inform better decision-making.

Education and Learning Theory. Educational psychology explores how intention influences learning processes and attentional focus in educational settings. Theories about goal-directed behavior highlight the importance of setting clear learning objectives and intentions to enhance focus and retention (Zimmerman, 2002). In an educational context, students who set specific learning intentions (e.g., mastering a particular concept) may fixate so much on the goal that they overlook broader contextual or related information, which could enrich their understanding. This highlights how intention, while guiding focus, can also limit awareness of other important educational stimuli.

Human-Computer Interaction (HCI). HCI studies how people interact with computers and technology, focusing on the design and use of interactive systems. This field examines how design elements guide user intention and attention during interactions with digital interfaces (Norman, 2013). In the context of HCI, the design choices made to facilitate user intentions can affect how users notice or fail to recognize information within interfaces. For example, if a user is intensely focused on completing a task on a software platform, they may overlook critical alerts or updates that do not align with their primary goal, leading to "not seeing" relevant information.

Mind-Body Medicine. Mind-body medicine examines how patients' thoughts, beliefs, and intentions can influence physical health and the body's responses. This field encompasses practices like meditation and visualization, which are believed to harness intention to promote well-being (Astin, Berman, & Wallace, 2003). When individuals cultivate specific health-related intentions, they often become more attuned to bodily cues and environmental signals. However, if individuals fixate solely on achieving a health goal (e.g., losing weight), they may ignore other health indicators or lifestyle changes that are equally important. This reflects how intention can narrow focus, leading to neglect of other relevant information in the quest for health-related outcomes.

Psycholinguistics. Psycholinguistics explores the intersection of language and cognitive processes, examining how intention influences language comprehension and production. This field looks at how individuals direct their attention when processing language and how that directs their understanding of context. In this context, speakers may have a specific intention when communicating, such as conveying a message, and may focus their attention on that intention while overlooking other nuances or contextual details that affect meaning and recognition. For example, a listener focused on a speaker's main point may miss non-verbal cues or contextual subtleties that alter the full understanding of the message.

Intention Concepts and Related Topics

We move now from "fields' to frameworks and concepts related to intention, with particular emphasis on their ties to neuroscience and their relevance to understanding why we may not see what is right in front of us. Each section includes an explanation, importance, a reference, and a summary of how that reference relates to the concept.

The topics related to intention briefly touched below include goal setting and future-oriented intentions, the intention-action gap, intention in decision-making, intentionality and consciousness, mindfulness, motor intentions and brain activity, and neural circuits of intention.

Goal Setting and Future-Oriented Intentions. Intention is a key component of goal-setting theory. Goal setting within psychological frameworks provides insight into how intentions shape behavior and attention over time. The act of setting goals involves activating cognitive resources to focus on specific future outcomes, directing one's attention and intention toward achieving those end states (Locke & Latham, 2002). The clarity of intention in goal setting can significantly impact attention allocation. When intentions are strongly tied to defined goals, individuals may become overly focused on achieving those goals, potentially resulting in missed opportunities for recognition of other relevant stimuli in their environment. For example, someone pursuing a career goal might overlook interpersonal dynamics or opportunities for learning that occur around them, indicating how future-oriented intentions can overshadow present-moment awareness.

Intention-Action Gap. The intention-action gap refers to the psychological and neurological discrepancy between the intention to perform a behavior and the actual execution of that behavior. This concept examines why individuals may fail to act on their intentions, involving both cognitive processes and neural pathways (Oettingen & Gollwitzer, 2010). This gap is relevant for understanding the limitations of conscious intent. Even when people are motivated to notice changes or engage with their environment, competing cognitive factors or neural distractions can lead to a failure to act on those intentions, resulting in not noticing what's directly in front of them.

Intention in Decision-Making (Behavioral Economics). In behavioral economics, the term "intention" is often used in the context of understanding how individuals form intentions based on their preferences and how these intentions influence their decision-making processes. Behavioral economists explore how stated intentions may not always align with actual behaviors, leading to discrepancies in expected outcomes (Thaler, 1991). This exploration is crucial for understanding how intentions shape attention. For instance, if a consumer intends to buy eco-friendly products but is distracted by marketing for cheaper, conventional options, their intention may not manifest in their purchasing behavior. This misalignment illustrates how focusing intentions can lead to overlooking other pertinent details.

Intentionality and Consciousness. Intentionality refers to the capacity of the mind to be directed toward something—essentially, the "aboutness" of mental states. In neuroscience, intentional states are often explored in relation to conscious awareness, examining how intentions can effectuate cognitive focus (Searle, 1983). This framework relates to the idea of "not seeing" in that when the mind is intentionally directed toward specific tasks, it can inhibit awareness of other relevant stimuli. The cognitive resources dedicated to fulfilling a particular intention can lead to blind spots, causing important changes to go unnoticed.

Mindfulness (Setting Intentions). In the context of mindfulness, the term "intention" is central to practices aimed at enhancing awareness and focus. Intention in mindfulness refers to the deliberate choice to focus on the present moment and to direct one's attention towards specific experiences, thoughts, or actions (Kabat-Zinn, 2005). Mindfulness practices help individuals cultivate clear intentions that can enhance their capacity to notice and acknowledge changes in their surroundings, thereby reducing the tendency not to see significant details. Set intentions in mindfulness can promote a heightened awareness that counters the blind spots typical when one's attention is scattered.

Motor Intentions and Brain Activity. Studies in neuroscience show that motor intentions—our plans to engage in movement—activate specific brain areas even before the execution of the action. Understanding how the brain prepares for these movements sheds light on how intention influences attention (Fried & Wilson, 2008). The activation patterns associated with motor intentions imply that the brain prioritizes certain actions based on our intentions, which can limit attentional resources available to perceive other stimuli. For instance, as a person gears up to respond to a situation (like catching a ball), their brain is focused on that action, which may cause them to miss other relevant events happening at that moment.

Neural Circuits of Intention. Intention in neuroscience refers to the neural circuits involved in goal-directed behavior. Specific brain regions—including the prefrontal cortex, parietal lobe, and basal ganglia—are known to play crucial roles in forming intentions and executing actions based on those intentions. These circuits regulate decision-making processes and motor actions based on desired outcomes (Cisek & Kalaska, 2010). The formation of intention is significant because when cognitive resources are focused on a specific goal, other stimuli may be overlooked. For example, if an individual is intensely focused on a task such as working on a report, they might not notice a colleague trying to get their attention. This highlights how the neural mechanisms of intention can lead to selective attention and, subsequently, missed opportunities for recognition.

Intention and Attention in Spirituality

Spirituality often emphasizes mindfulness, presence, and the intentional focus of attention to foster a deeper connection with oneself and the surrounding world. Many spiritual practices encourage individuals to cultivate awareness and direct their attention toward their experiences to achieve greater insight and understanding. This intentional focus is viewed as a pathway to enlightened perception, clarity, and spiritual growth.

Kabat-Zinn (2005) discusses how mindfulness practices enhance focus and awareness, advocating for the setting of intentions in daily life. He emphasizes that cultivating attention through mindfulness leads to an increased capacity to notice and appreciate life's nuances. This framework illustrates how spiritual practices can enhance perception, allowing individuals to recognize changes that might otherwise be overlooked. The book highlights how distractions or lack of intention can diminish awareness, contributing to the phenomenon of not seeing what is right in front of us.

In spiritual contexts, the cultivation of attention and intention can significantly impact how individuals perceive their environment. When practitioners engage in spiritual practices (such as meditation or prayer) with clear intentions, they develop heightened awareness of subtle changes or important messages in their surroundings. Conversely, if individuals are not mindful or are distracted, they may miss pivotal insights or experiences that are meant to be recognized. This understanding reflects the core spiritual theme of being present and fully engaged with the moment.

Let's specifically dig deeper into mindfulness meditation and spiritual intentionality.

Mindfulness meditation is a spiritual practice that fosters deep awareness and attentive presence. It encourages practitioners to focus their attention without judgment, cultivating intentions that align with self-awareness, compassion, and clarity. Through mindfulness meditation, individuals enhance their ability to direct attention to the present moment. This practice can help mitigate the cognitive overload that often results in missing important details. When practitioners intentionally focus on their breath or bodily sensations, they develop an acute awareness of their surroundings and may become more attuned to subtle changes in their environment.

Haris (2014) explores mindfulness and meditation as tools for enhancing attention and intention, discussing how these practices can lead to greater awareness of one's mental and emotional states while simultaneously heightening perception of external stimuli. This elevated state of awareness allows practitioners to recognize changes and insights that might otherwise go unnoticed, illustrating the vital role of spiritual practices in shaping attention and intention.

Scenario Example: In a busy office, Emma finds herself feeling overwhelmed as she juggles multiple projects and tight deadlines. Despite her best efforts, she often misses critical details in her reports, resulting in errors and frustration. Realizing the need for a change, Emma begins practicing mindfulness meditation every morning for ten minutes. During these sessions, she focuses intently on her breath, noticing each inhalation and exhalation with a non-judgmental awareness. As she develops this practice, Emma starts to carry the heightened sense of presence into her workday. One afternoon, during a tense meeting, she becomes aware of the subtle changes in her colleagues' body language and tone, information she had previously overlooked. This newfound awareness allows her to respond more thoughtfully to their concerns and contribute valuable insights. Over time, not only does Emma's work improve, but she also experiences reduced stress levels and a deeper connection to her surroundings, illustrating the transformative power of mindfulness meditation in helping her to see what was right in front of her all along.

Spiritual intentionality refers to the purposeful focus on spiritual goals, such as personal growth, enlightenment, or deeper connection with the universe. This involves not just setting intentions but aligning one's actions and attention with spiritual values and principles. When individuals set spiritual intentions, they often direct their attention toward practices that support these goals (e.g., prayer, meditation, acts of service). This focused intention can enhance their ability to see beyond everyday distractions, fostering a deeper understanding of their experiences and their connection to the world around them. However, if one's attention drifts from these intentions, they may become blind to vital spiritual insights or lessons in their immediate environment.

Palmer (1998) emphasizes the importance of intentionality in education and personal development, framing it within a spiritual context. He argues that when educators and individuals intentionally engage with their purpose, they can unlock deeper insights and connections. This focus on spiritual intentionality illustrates how aligning one's attention with spiritual goals can lead to a greater awareness of meaningful changes in the environment, enhancing perception and recognition.

Scenario Example: In her role as a teacher, Laura strives to not only educate her students academically but also inspire them to grow personally and spiritually. At the start of each school year, she sets a spiritual intention to cultivate compassion and empathy within her classroom. To align her actions with this intention, Laura incorporates practices such as beginning each day with a moment of silence, where students are encouraged to reflect on gratitude and positive aspirations. She also organizes community service projects that allow students to connect with and contribute to the wider world.

Over time, Laura notices a transformation. When a conflict arises between students, rather than reacting with anger or frustration, she takes a moment to breathe deeply and address the situation with understanding and patience, embodying the compassion she aims to teach. One day, during a class discussion, a student shares a personal story that deeply moves the entire class. It becomes a catalyst for a profound conversation about empathy and understanding diverse perspectives, a moment that reinforces Laura's intention.

Laura observes that her students not only develop better interpersonal skills but also become more attentive to each other's feelings, creating a supportive and harmonious environment. By adhering to her spiritual intention, Laura enhances her ability to sense and nurture meaningful connections, demonstrating the profound impact of spiritual intentionality on both her students and herself.

As can be seen, the field of spirituality connects deeply with the concepts of attention and intention, revealing how these elements enhance perception and awareness. Through practices such as mindfulness meditation, spiritual intentionality, and focused awareness, individuals can cultivate the ability to notice significant changes and insights in their environment. This exploration emphasizes the interplay between spirituality and cognitive processes, highlighting how intentional focus can help individuals counter the tendency not to see what is right in front of them.

After Thoughts

Intention and attention are intricately linked cognitive processes that work together to shape human behavior and experience. Intention serves as the motivational framework that guides our actions and choices, indicating what we want to achieve or accomplish. It provides the purpose behind our activities, whether those are long-term goals, like pursuing a career, or short-term actions, like focusing on a conversation. Attention, on the other hand, is the cognitive mechanism that allows us to filter and prioritize information from our environment, enabling us to concentrate on what is relevant and significants.

When we set an intention, it helps to direct our attention toward specific stimuli or tasks that align with our goals. For instance, if someone intends to exercise regularly, their attention will likely be drawn to opportunities for physical activity, such as gym facilities, fitness classes, or even casual reminders in their daily environment. This focused attention ensures that they remain cognizant of their intention, facilitating the cognitive and behavioral effort needed to turn that intention into action. Conversely, when attention is scattered or unfocused, it can hinder the execution of intentions. For example, if distractions from the surrounding environment pull focus away from a task (like studying for an exam), it becomes challenging for the individual to stay committed to their intention of learning.

Moreover, the interplay between intention and attention can influence the outcomes of behavior and decision-making. Research has shown that when individuals have clear intentions and are able to maintain focused attention on related activities, they are more likely to experience positive results and a heightened sense of accomplishment. This synergy not only fosters greater performance in tasks, but also reinforces motivation, as successful alignment of intention and attention can bolster confidence and satisfaction. Thus, understanding how these two cognitive processes work together can provide deeper insights into human behavior, enhancing our capabilities to achieve goals and optimize performance in various aspects of life. Together, they set the stage for creating our reality.

Chapter 8
Relativism

Amidst views where relativism plays,
Truth takes on multifaceted ways.
With morals and taste,
We *Whysly* retrace
Paths where diversity holds sway.

When thinking about our core topic (why we can't see what is right in front of us), there was no question in my mind of the importance of relativity. Yes, I reflected on the term "relativity" because my life partner of 30 years was a nuclear physicist/neuroscientist and relativity, well, just was almost a common household word. So I set about making sure I understood and could explain relativity and its relationship to our focus.

In physics, "relativity" stems from Albert Einstein's theories—special and general relativity—which revolutionized our understanding of space, time, and gravity. It describes the behavior of objects in motion and how gravity affects the shape of space and the flow of time. In relativity, physical laws and observations can vary based on the observer's frame of reference, but these laws are universally valid when expressed correctly. It's a scientific model, with precise mathematical formulations and testable predictions.

The philosophical concept of "relativism" addresses the nature of truth and morality, suggesting that beliefs and values are contingent on cultural, historical, or individual contexts. Relativism argues that there are no absolute truths or moral standards that apply universally; instead, truth is always relative to some particular framework of understanding.

Philosophers sometimes use relativity as a metaphor to illustrate how perspective can shape interpretations of truth and morality, much like how observations in physics are frame-dependent. However, the connection is metaphorical rather than scientific. Relativity does not make philosophical claims about truth or ethics, whereas relativism does not provide mathematical models or predict physical phenomena. In essence, while both concepts emphasize the importance of perspective, relativity is concerned with physical observations and laws, while relativism addresses human perspectives on truth and morality. They operate independently in their respective domains.

Historical Development of Relativism

The philosophical concept of relativism—upholding the idea that points of view have no absolute truth or validity, but are instead relative to the perspectives or contexts from which they arise—has evolved over time, drawing from various cultural and intellectual traditions. The philosophical roots can be traced back to ancient Greece, particularly to the Sophists (5th Century BCE), a group of itinerant teachers who argued that truth is relative to the beliefs and perceptions of individuals. Protagoras, one of the most famous Sophists, famously stated, *Man is the measure of all things* (Guthrie, 1971). Plato and Aristotle both responded critically to relativism, arguing for universal truths and forms (in Plato's case) (Plato, 2003) and natural law (in Aristotle's case) (Aristotle, 2009). However, medieval philosophers like Thomas Aquinas maintained that truths about God and ethics were universal, rejecting relativism (Aquinas, 1911-1925/1265-1274).

In the 19th Century, Friedrich Nietzsche is often linked to relativism, especially in rejecting objective truth and promoting perspectivism (the idea that all knowledge is interpreted through a lens, foundational to the field of Knowledge Management. Nietzsche's perspectivism suggested that even science is an interpretation among many, not a final or universal truth (Nietzsche 1989/1887).

In anthropology, Franz Boas (1940) challenged the ethnocentric views prevalent in the 19th century by advocating for cultural relativism, the idea that a person's beliefs and activities should be understood based on that individual's own culture. This shifted the study of cultures from a hierarchical framework to one of appreciation of diversity, a lesson that has often been forgotten in today's societies. Interestingly, while Edmund Husserl (1970/1936) and phenomenology argued against total relativism, they acknowledged the role of cultural-historical context in shaping meaning.

In *The Structure of Scientific Revolutions*, Thomas Kuhn (1962) proposed that scientific truth is relative to paradigms, that is, what counts as evidence or explanation depends on the scientific consensus of the time. He stops short of full-blow relativism. Not too many years later, in *Philosophy and the Mirror of Nature,* Richard Rorty (1979) espoused a version of epistemic relativism, arguing that truth is a product of linguistic frameworks and no framework is privileged. Both the knowledge movement and, more recently, the emergence of Whole Thought as a cognitive framework embrace epistemic relativism (Bennet & Turner, 2023).

Relativism has had profound effects in ethics where it is a major topic in debates about multiculturalism, human rights, and social reform. In epistemology, relativism challenges the quest for universal foundations. In political philosophy, it interacts with liberal ideas of tolerance and pluralism. It is also important in the social sciences (especially anthropology), literary theory, and critiques of Western-centric or universalist narratives.

Let's briefly explore areas of research where relativism has become heavily rooted, beginning with moral relativism, cultural relativism, and epistemological relativism.

Moral, Cultural, and Epistemological Relativism

At its core, **moral relativism** is the meta-ethical position that the truth (or justification) of moral judgments is not objective or absolute but is relative to some standpoint—most commonly, a culture (cultural relativism) or an individual (subjectivism). Unlike conventionalism or pluralism, genuine moral relativism claims there is no *neutral*, universal standpoint from which to adjudicate all moral claims.

There are three distinguishing "types" of cultural relativism: descriptive, meta-ethical moral, and normative ethical relativism. *Descriptive moral relativism* is an empirical claim. Different cultures/individuals hold divergent moral beliefs. *Meta-ethical moral relativism* says the truth/falsity of moral claims is relative to the speaker's or assessor's moral framework. *Normative ethical relativism* is the view that what is morally right or wrong depends on the cultural, societal, or individual contexts such that ethical truths vary from one society or individual to another.

As documented by anthropologists, there is consistent disagreement that exists between and within this domain of research, which prevents attempts at cross-cultural conversations (Harman, 1975; Wong, 1984).

Cultural relativism is the view that beliefs, values, and practices are properly understood *only* within the context of a given culture. In ethics, it is the claim that what is right or wrong must be judged according to local cultural norms, rather than by any "universal" standard.

Similar to moral relativism, there are three distinguishing "types" of cultural relativism. These are methodological, descriptive, and normative (meta-ethical). *Methodological cultural relativism* is a method for understanding and studying cultures; all beliefs and practices should be understood in context, suspending one's own biases. *Descriptive cultural relativism* is the empirical observation that moral codes and social norms actually vary from culture to culture. *Normative (meta-ethical) cultural relativism* embraces the observation that moral codes and social norms actually vary from culture to culture.

Cultural relativism has been a powerful force in anthropology, moral philosophy, and politics. It brought the invaluable lesson that human diversity is deep and often non-comparable, and it challenges the arrogance of ethnocentric standards. Yet it faces ongoing difficulties—above all, how to justify inter-cultural criticism, how to avoid moral paralysis in the face of atrocity, and whether it can defend its own principle of tolerance without appealing to some values as universal.

Epistemological relativism is the view that knowledge, justification, rationality, or truth are not absolute or universal, but relative to frameworks, cultures, paradigms, or individual perspectives. The core claim could be described in this manner: There is no single, universally binding set of standards for evaluating knowledge or justification claims. What counts as *knowledge*, *rational belief*, or *adequate justification* can only be settled *relative to* specific cultures, historical periods, paradigms, social practices, or languages. Note that this is based on the philosophical description of knowledge as "justified true belief" or, in today's vernacular, "the capacity to take effective action". A core argument against this is that if there is no absolute truth, then this statement itself becomes an "absolute," a paradox.

In the field of Knowledge Management, knowledge is considered context-sensitive and situation-dependent. When knowledge is explicit, it is easily pulled into the conscious mind and communicated. When knowledge is tacit, it resides in the unconscious and is not easily if even possible to communicate. Thus, we add the relevance of the decision-maker to the context and situation where the decision is being made. Decisions are also temporally relevant, adding a fourth dimension to the knowledge (truth) of the moment.

Here is a comparative chart of moral relativism, cultural relativism, and epistemological relativism.

Aspect	Moral Relativism	Cultural Relativism	Epistemological Relativism
Definition	Truth or justification of moral judgments is relative to a moral framework, culture, or individual.	Beliefs, values, and practices are intelligible or valid only in the context of a particular culture; often, moral standards are seen as culture-bound.	Justification, rationality, or knowledge are relative to a paradigm, framework, culture, or social group.
Object of Relativity	Culture/framework/ individual	Culture/society	Epistemic system/ paradigm

Central Claim	**No universal moral truths.**	**All norms/beliefs are context-bound.**	**No universal rational/ justification standards.** In KM: All knowledge (truth as knowledge) is context-sensitive, situation-dependent, temporally and individually relevant.
Entry Points	Meta-ethics, value pluralism, applied ethics	Anthropology, moral philosophy, legal and rights discussions	Philosophy of science, knowledge debates, sociology of science
Paradigm Example	"Stealing is wrong" is true/false relative to society	"Wearing shoes indoors is wrong"—varies by culture	"What counts as science?" varies across paradigms

Importance to Our Focus

Our ability to recognize and respond to truths or problems directly before us is often undermined by the frameworks through which we interpret the world. Relativism's overall connection to the theme of not seeing what is right in front of us lies in its assertion that *perception is deeply subjective*. When one's viewpoint is steeped in a specific cultural or moral context, this may obscure objective realities or competing perspectives. The relativistic lens might lead individuals to prioritize localized truths over universal truths, blinding them to broader, more objective observations.

Moral relativism posits that moral judgments and values are not universal but instead shaped by cultural, social, or personal perspectives. This perspective can obscure what's right in front of us by making it challenging to discern any one moral truth or action as clearly "right" or "wrong." When our moral compass is contingent on context, consensus, or upbringing, even the most obvious ethical dilemmas can become ambiguous or invisible. For example, a practice that seems unquestionably unjust to outsiders—such as systemic discrimination—may go unchallenged within a community if it's normalized by their relative ethical standards. Thus, moral relativism helps explain why we may fail to recognize—or feel compelled to address—practices that are unjust right before our eyes.

Cultural relativism is the view that beliefs, values, and practices should be understood within their own cultural context, without judging them by external standards. This approach, while encouraging tolerance and understanding, may also blind us to truths that seem self-evident from a broader or external perspective. If we continually defer to "that's just their way" or "it's part of the culture," or even "politicians always lie, so what?" we risk accepting harmful or questionable norms simply because they are traditional or locally accepted. This can prevent us from questioning or reevaluating practices and assumptions within our own society, causing us to overlook injustices or errors that are, figuratively and literally, right in front of us. Cultural relativism thus contributes to a form of collective myopia, inhibiting our ability to see and act upon problems that are culturally embedded.

Epistemological relativism holds that knowledge and truth are not absolute but are instead contingent upon various frameworks—such as language, conceptual schemes, or cultural paradigms. Under this view, even evidence or facts that seem self-evident can be dismissed or reinterpreted depending on one's epistemic standpoint. This can have profound implications for why we fail to see what's in front of us: confirmation bias, worldview, or linguistic constraints may all filter or shape our understanding of reality, rendering us unable to recognize clear evidence or obvious truths. Epistemological relativism thus highlights how our basic modes of knowing can obscure apparent facts, trapping us within limited perspectives and preventing us from seeing reality as it actually is.

Scenario Example: Consider Emma, an anthropologist raised in a Western context, researching a remote society that values communal decision-making over individualism. Through her lens, Emma initially views their practices as inefficient. However, over time, as she immerses herself in their culture, she begins to question her assumptions and sees the strengths in their approach. Initially, Emma's cultural relativism caused her to overlook the effectiveness and satisfaction the community experienced through their system. Her journey illustrates the challenge of overcoming ingrained cultural narratives to truly perceive what lies in view but was initially obscured by her own cultural biases.

In summary, moral, cultural, and epistemological relativism each contribute to the phenomenon of being unable to see what is right in front of us by challenging he idea of objective truth or reality. Moral relativism blurs ethical clarity, making it difficult to recognize injustice or wrongdoing when it is normalized within a particular context. Cultural relativism fosters tolerance but often at the expense of critical reflection, encouraging acceptance of practices simply because they are culturally sanctioned. Epistemological relativism, meanwhile, questions the very foundation of knowledge, making it possible for individuals or groups to dismiss or reinterpret clear evidence. Together, these forms of relativism reveal how the interpretive frameworks shaping our perceptions can obscure the obvious, explaining why even the most apparent truths or problems can become invisible to those immersed within particular moral, cultural, or epistemic standpoints.

Although not as well-known, there are other types of relativism. We will briefly explore a few of these pertinent to our focus. These include aesthetic (or taste), legal, linguistic (or semantic), logical, and science or technology studies (STS) relativism.

Aesthetic (or Taste) Relativism

Aesthetics is a key area in the literature *on semantic* or *assessment-sensitive relativism* (MacFarlane, 2014). It is especially robust in philosophy and cognitive science, parallel to moral or epistemic relativism in scope and depth. MacFarlane introduces the idea of "assessment-sensitivity," clarifying how statements can be evaluated as true or false not just relative to context of utterance, but to *context of assessment*. This work is key to bringing semantic clarity and rigor to relativism in aesthetics (as well as epistemology and morality).

The core idea of aesthetic relativism is the recognition that judgments about beauty, art, and taste are true or appropriate only relative to individual sensibilities, group standards, or cultural frameworks. David Hume's (1757/2008) *Of the Standard of Taste* struggles with whether any universal standard for artistic value is possible. He argues that while the sentiments of beauty vary, there are nevertheless some broad shared standards, albeit always moderated by individual and cultural context. His essay is a foundational work in the philosophy of art, raising enduring problems for both relativists and their critics regarding the universality of aesthetic judgment.

The idea of "faultless disagreement" emerges in aesthetics (Kölbel, 2004). This is when two people can disagree about taste, a painting or music with neither being "wrong." This view undergirds much recent semantic relativist work, supporting the notion that aesthetic claims are true or false only relative to standards not universally shared. This directly parallels debates in moral relativism. Examples are "cilantro tastes bad" or "this painting is beautiful," both statements whose truth is indexed to individual or cultural standards. "Faultless disagreement" is an excellent approach to apply in social relationships.

Scenario Example: Two friends, Alice and Ben, visit a modern art gallery. Alice is moved to tears by an abstract sculpture, declaring it a masterpiece of emotional depth. Ben, meanwhile, is unimpressed and thinks it's meaningless. They argue, but neither can convince the other—even the gallery guide tells them that "beauty is in the eye of the beholder." This scenario shows *faultless disagreement* and relativism about taste: Alice's and Ben's judgments are both correct, but only relative to each person's aesthetic sensibility and evaluative standards.

Connection to Our Focus: Neither Alice nor Ben recognizes that their reactions aren't *universally* compelling but are embedded in personal history, taste, and perhaps cultural influences. What's "right in front" of Ben—a lack of emotional impact—is opaque to Alice, for whom beauty is radiant. Their own standards are invisible to them; they assume beauty is (or should be) apparent to all. Our standards for beauty and art are so engrained they vanish from view; each person feels they're just seeing what is real, not through their own lens.

Legal Relativism

Used in anthropology, comparative law, and legal philosophy, the core idea of legal relativism is that concepts of legal justice, right, and even basic elements of law (property, marriage, contract) vary fundamentally across societies. Pound (1912) was a pioneer in legal relativism, focusing on the social context of law, arguing that legal norms and decisions should be understood in relation to the societies in which they operate—a position that underpins much later work on legal relativism.

Tamanaha (2000) demonstrates the contextual, culture-relative nature of legal systems and norms, challenging legal universalism. Critiques of "universal law" frameworks (as with human rights) often invoke legal relativism. For example, marriage law, ideas of property, and due process differ radically worldwide. Merry (1988) explored the persistence of multiple coexisting and conflicting legal systems *within* societies, especially in colonial and postcolonial contexts, as well as the interplay between state law and customary or religious practices. This work is a touchstone for legal and anthropological relativism.

Scenario Example: A land dispute in rural India is brought before a local panchayat (village council), which uses customary law: family, caste, oral agreements, and tradition govern the outcome. Yet the region is also subject to national law, which follows British-influenced legal codes. The same dispute might be decided differently depending on which system the parties choose—and both systems are seen as legitimate by locals. In this example, legal pluralism and relativism are evident: the legitimacy and very *concept* of law varies across overlapping systems, and resolution depends on the legal tradition to which appeal is made.

Connection to Our Focus: Villagers and legal professionals may not recognize their own legal system's arbitrariness; "justice" seems self-evident in their procedural and cultural terms. Only when the clash between legal systems emerges does it become clear that different forms of "justice" can be simultaneously present, yet mutually invisible. Our sense of "law" or "justice" feels natural and simply "right"—making it surprisingly hard to see even radically different systems functioning in the same community.

Linguistic (or Semantic) Relativism

Linguistic relativism is robust in philosophy and cognitive science. Linguistic relativism, also known as the Sapir-Whorf hypothesis, is the idea that the language we speak influences how we think and perceive the world. Meaning, truth, and even the way we experience reality are relative to language or conceptual schemes. For example, speakers of different languages may have different ways of describing color, potentially affecting their perception and categorization of colors.

Benjamin Lee Whorf (1956), in his collection *Language, Thought, and Reality*, argued that language shapes thought and perception, a principle that underlies the theory of linguistic relativity. Whorf's studies, such as his research on the Hopi language, suggested that linguistic structures and vocabularies could significantly impact cognitive processes by framing a speaker's experience of space, time, and reality in ways that differ from speakers of other languages. In other words, linguistic structure shape how speakers experience reality. Whorf's work opened the door to exploring how language's influence might extend beyond communication, affecting cognitive categories and perceptual distinctions.

Linguistic relativity is pertinent to understanding why we sometimes cannot see what's right in front of us in a metaphorical sense. Language not only reflects but also shapes our perceptual categories, potentially blinding us to certain details in our environment that fall outside these linguistic frameworks. When a particular language emphasizes specific concepts or categories, it may lead speakers to prioritize these aspects, overlooking others. This cognitive framing illuminates how cultural and linguistic factors might inadvertently filter reality, making certain things less noticeable despite their presence. Thus, linguistic relativity offers insights into how our language-conditioned mindset could contribute to an absence of awareness or attention.

Scenario Example: On a language immersion program in Papua New Guinea, linguist Jade tries to translate the English word "time." She discovers the local language has no tense system and speaks instead in terms of seasons and events, not a past-present-future continuum. When she talks about "wasting time," her hosts are puzzled: time can't be wasted, only experienced. Jade's experience reflects linguistic relativity. The concepts she sees as basic don't exist in the same way in her hosts' language, showing how language deeply shapes conceptual schemes.

Connection to Our Focus: To Jade, "time" is an obvious, ever-present feature of reality; the idea that someone could *not* have this concept never even occurs to her. Only through difficult cross-cultural encounters does she realize what is "right in front of her" is structured by her language—a lens so transparent she seldom sees it. Similarly, her hosts' world is organized so differently that her concept of "wasting time" is literally unthinkable to them. In both cases, our language and assumptions form a "conceptual apparatus" that makes alternatives invisible; the most basic elements of our reality are seen *only* through their mediation.

Logical Relativism

The laws of logic (e.g., law of excluded middle, principle of non-contradiction) might themselves be local to logical systems, with no "absolute" logic. The law of excluded middle states that for any proposition, either that proposition is true or its negation is true. In formal terms, given a proposition (P), the law can be expressed as (P/|or/|negP where (/|or) represents "or" and (/|neg) represents "not"). This means that there are no other truth values possible aside from either being true or false. The principles of non-contradiction assert that contradictory propositions cannot both be true at the same time. In formal terms, it is expressed as (/neg (P/|and /neg P)) where (/|and) represents "and"). This means that a statement and its negation cannot both hold true.

There are still significant debates about whether classical logic is uniquely correct, or one among several other approaches such as paraconsistent logic (allowing for contradictory statements to coexist without leading to triviality, meaning not all statements can be derived as true from the contradictions) and intuitionistic logic (emphasizing constructivist philosophy, asserting that the truth of a proposition is tied to our ability to prove it, thus rejecting the law of excluded middle).

Beall and Restall (2006) argue that more than one logic can be correct, depending on the legitimate standards or contexts considered ("logical pluralism"). Their work challenges the notion of a single, context-independent logic, supporting a kind of relativism within formal reasoning itself. Technically a part of analytic philosophy, especially the philosophy of logic, the notion of "logical pluralism" is closely related. Logical pluralism is the view that there are multiple legitimate logics that can be applied to different contexts, allowing for diverse interpretations of truth and validity. Field (2009) investigates logical paradoxes and the foundations of logic, arguing that solutions may require employing non-classical logics. His exploration of "logical pluralism" and context-dependent logic contributes to the discussion of logical relativism.

Reflect: Classical logic says: "Every proposition is either true or false." Intuitionistic logic says: "a proposition is true only if we can prove it." In your opinion, is one of these *the* correct logic?

Scenario Example: An undergraduate class debates whether certain logical paradoxes (like "This sentence is false") are truly contradictions. Some students prefer intuitionistic logic, which doesn't allow proof of some classical laws (like excluded middle), while others stick to classical logic. Their instructor points out both logics are internally consistent, but which is "right" depends on context and purpose. Here, logical relativism is on display: differing logical systems are valid in their own rights, and which set of logical laws to adopt depends on underlying standards and goals.

Connection to Our Theme: To each student, the rules of logic are so foundational they seem to define what it means to "think straight," making the other's logic seem alien or nonsensical—each side blind to the contingency of their "obvious" starting points. The laws of logic become so deeply internalized they're no longer seen as *adopted* or *chosen*; rival logics can become literally unseeable or "unthinkable" to the initiates of a tradition.

Science and Technology Studies (STS) Relativism

Grounded in the sociology of science and technology, the core idea of STS is that fact and technology construction are contingent on social, historical, and cultural forces—scientific "truths" are not purely objective but partly relative to scientific cultures. My 30-year nuclear physicist and neuroscientist partner questionned this concept.

Kuhn's (1962/1970) landmark study introduced the idea of "paradigm shifts," arguing that scientific understanding is not linear or cumulative but periodically undergoes radical transformations that change standards of evidence, explanation, and rationality. While Kuhn resisted strong relativism, his work is foundational to the idea that standards of rationality and "truth" in science are paradigm-relative

There is some lively debate about constructivism versus realism in the philosophy of science. Bloor (1976/1991) argues the strong relativity thesis, that all knowledge (including scientific and mathematical) is culturally local. Latour and Woolgar (1979/1986) argue *methodological relativism*, the construction of facts. This is tightly linked to epistemological relativism.

Scenario Example: A team of physicists debate the evidence for dark matter. Older scientists, trained before recent breakthroughs, dismiss certain experimental anomalies, while younger postdocs interpret them as groundbreaking. Both sides use complex statistical reasoning—each based on the methods and assumptions standard in their formative years. Even methods themselves are debated. This scenario exemplifies Kuhnian paradigm-relativity in science: evidence and rational inference are not purely objective, but are shaped by the "paradigm" or tradition that scientists operate within.

Connection to Our Focus: For each camp, the standards of scientific evidence and reasoning are not up for question—they are so embedded that other approaches seem irrational or even invisible. Paradigms act as "blinkers," limiting what data is noticed or how it is interpreted. Scientific paradigms are not just tools—they shape perception itself, so that alternatives are not just wrong, but often literally unseeable, even when "in plain sight."

Recognizing Relativism in Yourself

Is there a way to break away from relativism, or at least recognize it in yourself? That's an important and enduring question—central to both philosophy and self-understanding. The nuanced answer below blends philosophical argument, psychological insight, and practical suggestions.

There are *limits of transcendence*. Philosophically, the strongest anti-relativists (e.g., moral realists, epistemic objectivists) argue that self-reflection, reasoning, and universal dialogue can reveal truths transcending any particular framework. Yet, others argue that (1) you always reason from some framework, even if you switch or expand frameworks (Quine, 1951; Kune, 1962/1970); (2) we must have enough shared ground to communicate, suggesting there is at least partial escape from total relativism—but total transcendence of all frameworks may be possible (Davidson, 1974); and (3) we interrogate frameworks by "making them objects of comparison"—but the act of comparison itself is always rooted somewhere (Winch, 1958; Taylor, 1985). In short, it's very hard—maybe impossible—to "see" entirely outside all frameworks *simultaneously*; but it is very possible (and valuable) to recognize, modify, and critically distance yourself from *particular* frameworks.

The first step is recognition, becoming aware that what you take as "obvious" or "natural" (in taste, knowledge, or morality) is likely framed by your background, culture, socialization, and cognitive "habits"—not by objective necessity. This is the goal of much of the Socratic phenomenological method, that is, unmasking hidden, unexamined assumptions. (See Appendix

A.) Socrates called this "examined life." Heidegger called such frameworks our "pre-understanding."

Practices that foster awareness include cross-cultural encounters, dialog with difference, and the study of anthropology, history, and philosophy. Exposure—ideally immersive—to radically different cultures, traditions, and worldviews is perhaps the surest way many people come to feel the relativity of their assumptions. Exposure—ideally immersive—to radically different cultures, traditions, and worldviews is perhaps the surest way many people come to feel the relativity of their assumptions. Seeking out genuine dialog partners from other backgrounds and listening deeply (not for debate but for understanding) helps uncover your own blind spots. Learning how other times, places, and thinkers have organized their lives, knowledge, and values "denaturalizes" your own framework.

Here's some practical steps toward transcendence with a scenario example.

Context of Scenario: Allie is a project manager at a multinational company. She's leading a team that includes colleagues from diverse cultural backgrounds, including Japan, India, and Brazil.

Step 1: Reflexivity. Routinely ask: *What assumptions am I making? Where did they come from? How do others do or see things differently?*

Scenario Example: During a meeting discussing project timelines, she notices that her instinct is to prioritize efficiency and time management, values deeply rooted in her Western work culture. Allie decides to pause and reflect: What assumptions am I making about how projects should be managed? She realizes her preference for direct communication and rigid timelines might not be shared universally.

Step 2: Philosophical Method. Use reason, argument, and moral imagination to critically appraise not just other systems but your own—a key move in anti-relativist arguments.

Scenario Example: To explore this further, Emma initiates an inclusive discussion during the next meeting, encouraging team members to share their own cultural approaches to project management. She uses reason and argument to unpack the pros and cons of different methods, examining her own system critically by comparing it against others without bias. She discovers that while her method prioritizes speed, others emphasize quality or team harmony, each with their own rationale and moral standing.

Step 3: Imaginative Empathy. Exercise the capacity to imagine *from within* another way of being, knowing, or valuing. *Example:* Try "moral experiments" imagining you were raised in a different culture, with different knowledge traditions.

Scenario Example: Allie practices imaginative empathy by conducting a "moral experiment." She envisions herself growing up in Japan, where indirect communication and consensus are valued over speed in decision-making. By imagining herself in her colleague Hiroshi's shoes, she begins to see the benefits of patience and collective agreement. This exercise helps her appreciate the perspective that prioritizes relationship-building over immediate results.

Step 4: Fallibilism. Adopt the attitude that your framework might be mistaken or incomplete. *Open to humility*. This openness guards against dogmatism (and relativistic complacency) and supports reasoned dialogue.

Scenario Example: Allie embraces fallibilism by acknowledging that her approach may be incomplete or unsuitable in all contexts. She proposes a hybrid strategy for the project: setting flexible timelines that allow for team input and periodic checkpoints to ensure quality and consensus. She maintains an open dialogue, encouraging team members to voice discomfort or suggestions.

Through this process, Allie not only develops a more effective project strategy but also grows in her understanding of different cultural frameworks. She recognizes that her initial assumptions were just one way of seeing things, fostering her leadership skills and enhancing team cohesion and respect.

Below is a summary table.

Step	What You Do	Outcome
1. Recognition	Notice your own background assumptions	You see they are not "necessary"
2. Comparison	Encounter other frameworks (cultures, beliefs)	Your own becomes visible/contestable
3. Reflexivity	Subject your beliefs to critical examination	Begin to "see differently"
4. Deliberation	Seek rational grounds across differences	Possible reformation/transcendence

What Breaking Away Is, and Isn't

You may never escape all frameworks. Complete "objectivity" is arguably an illusion. BUT ...

- You can *move* between frameworks.
- You can *see* the limitations and contingency of your own.
- You can *seek common ground* (reason, empathy, universals) that enables real critique, dialogue, reform, and learning.
- You can use "framework-awareness" as a tool, not a trap.

On a philosophical note, anti-relativist positions (e.g., those of moral realists) (Boghossian, 2006) rest on the belief that, through philosophy, science, and intersubjective *deliberation, justification or truth* can be established on grounds not hostage to arbitrary perspectives. Even if you accept some relativity, you can—by critical reflection, encounter, and reason—rise above many (if not all) of your parochial limits. When you stretch the scope of your understanding, judgment, and empathy, the result is not "now I'm objective" but a humility, openness, and critical capacity that allows you to see more of what's right in front of you, and see it more deeply.

Chapter 9
Blindness

In a world where blindness does bind,
Each type leaves a shadow on mind.
Through *Whysly* pursuit,
We seek to uproot
These veils, for the clarity we find.

In the movie "Hacksaw Ridge," a wounded soldier lay on the battlefield in darkness both startled and thankful when a medic discovers him and quickly administers a pain killer. "I can't see" he exclaims. The medic quickly treats his eyes, pouring water over them to clear the mud away. I can only imagine the elation in that instant of first sight, and it brings to mind a line from the hymn *Amazing Grace*: "I was blind but now I see."

As the value of knowledge moved into our awareness at the turn of the century, we learned that *we don't know what we don't know*. Similarly, *we can't see what we cannot see*.

There are many forms of blindness that can prevent us from seeing that which is right in front of us. These include the physical and mental processing explored in the previous chapter as well as change, cultural, economical, emotional, ideological, inattentional, moral, perceptual, political, psychological, social, and even temporal blindness. While we relate each of these forms with the primary field(s) they come out of, you will begin to see patterns of an integrated understanding emerging.

Change Blindness (Cognitive Psychology/Perception)

Change blindness refers to the phenomenon where individuals fail to notice significant changes in a visual scene when the change occurs during a visual disruption, such as a flicker or a brief occlusion. This inability to detect changes highlights the limitations of visual perception and attention, underscoring that our perception of the world is not as seamless as we may believe. For example, in a well-known demonstration of change blindness, people are shown images that undergo slight modifications while their attention is focused elsewhere (e.g., during a brief interruption or a cut from one scene to another). Participants often fail to notice these changes, illustrating the limits of their visual attention.

Research on change blindness, especially by Simons and Levin (1997), directly answers why dramatic changes can go undetected even when they happen before our eyes. Studies show that unless attention is explicitly drawn to a change, observers' mental representations fill in the gaps, maintaining continuity even in the face of contradiction. The phenomenon reveals that our sense of a complete, accurate visual world is an illusion—our perception depends on attention, and without it, we're blind to even the most obvious changes right in front of us

Scenario: Imagine a storefront that undergoes a substantial visual redesign. A team conducting a market analysis visits the store and focuses on customer reactions to a new product display while missing significant alterations in the store layout itself. This oversight demonstrates how their focus on one aspect (customer behavior) led to a lack of awareness of another key element (the layout change).

Let's explore several applications in the organizational context. Change blindness can apply to organizations in various ways:

Employee Training: A company might implement new software that drastically alters workflow. If employees are not made aware of these changes beforehand or given adequate training, they may fail to recognize how the new system affects their daily tasks. *Scenario:* A company announces updated security procedures involving changes to access points within the facility. However, when employees are engaged in their routine tasks, many fail to notice the new security protocols or changes to entrance points, leading to confusion and security risks.

Customer Feedback: A marketing team might conduct surveys while neglecting to observe or inquire about changes in customer preferences or market trends. The team may present new product lines based on previous feedback without realizing that consumer desires have shifted. *Scenario:* A clothing retailer focuses solely on customer satisfaction ratings for existing styles, failing to investigate changes in fashion trends. When they launch new inventory reflecting outdated styles, they receive minimal customer interest due to their failure to notice shifting preferences.

Organizational Change: A clothing retailer focuses solely on customer satisfaction ratings for existing styles, failing to investigate changes in fashion trends. When they launch new inventory reflecting outdated styles, they receive minimal customer interest due to their failure to notice shifting preferences. *Scenario:* After a merger, employees continue to operate under the assumption that organizational culture remains unchanged, missing altered processes and communication styles brought in by the new leadership team. This lack of awareness can lead to misunderstandings and disengagement.

Cultural Blindness (Cultural Studies/Sociology)

Cultural blindness is the inability to recognize or appreciate cultural differences, resulting in a narrowed worldview. Such blindness can prevent individuals from understanding diverse perspectives, leading to stereotypes and cultural insensitivity (Hofstede, 1980). For example, An international business negotiator assumes that all partners share the same values and negotiation styles, failing to account for cultural preferences in communication or decision-making.

Studies inspired by Hofstede's cultural dimensions demonstrate that cultural cues are often so embedded in the background of social situations that those unfamiliar with a particular culture simply don't "see" them—even when they are explicit to insiders. Research indicates that cultural scripts and norms operate below the level of conscious awareness unless actively studied or brought to attention. Thus, "not seeing what's right in front of me" is frequently about an absence of the frameworks needed to decode meaning in cross-cultural situations, rendering critical nuances and perspectives functionally invisible

Scenario: During a negotiation with a Japanese company, the negotiator's direct approach is misinterpreted as rude, causing mistrust and tension. Acknowledging cultural norms could have fostered a more collaborative environment.

Economic Blindness (Economics)

Economic blindness refers to the inability to perceive financial realities or the broader economic forces at play due to inherent biases, lack of knowledge, or seeing only within the scope of one's immediate economic experiences. This type of blindness can manifest in various ways, such as underestimating the impact of socioeconomic factors on personal or societal levels, or failing to recognize economic opportunities and threats. It involves a narrow focus on personal financial situations without understanding how larger market forces and economic trends influence them. For example, a person might not notice the gradual rise in cost of living due to inflation because they only focus on their unchanged monthly income.

Economist John Kenneth Galbraith (1958) critiques the conventional economic assumptions of his time, particularly the focus on production and economic growth *at the expense of addressing public needs and social welfare.* He introduces the concept of "social balance," emphasizing that an increase in private wealth often leads to public squalor if societal issues like infrastructure, education, and poverty are neglected. Galbraith's work highlights the economic blindness of ignoring these realities, advocating for policy that recognizes and addresses the broader social impacts of economic growth.

Related to our focus, economic blindness affects our ability to see what's right in front of us by limiting our understanding of how economic factors impact our lives. Individuals may fail to perceive how macroeconomic trends, such as unemployment rates or fiscal policies, affect their personal situations. This blindness can lead to misconceptions about the stability of their economic environment or missed opportunities for improvement and adaptation to changes. By only focusing on immediate financial concerns, people often overlook the larger economic picture that shapes their daily realities.

Scenario Example: Consider a worker who is unaware of the technological advancements increasing productivity across their industry due to economic blindness. This worker continues to complete tasks using outdated methods, not realizing their job security is at risk because they can't see the shifting economic factors encouraging automation. Despite the growing trend of automation being right in front of them via news reports and industry publications, their focus remains on day-to-day performance, not recognizing how these developments will eventually influence their employment stability.

Emotional Blindness (Psychology)

Emotional blindness refers to the inability to perceive or understand emotions in oneself or others. It often manifests as a detachment from one's feelings and a lack of empathy toward others. This form of blindness hampers emotional intelligence, making it difficult to effectively navigate social interactions and respond appropriately to emotional cues. For example, a manager fails to notice that a team member is struggling with a heavy workload, missing non-verbal cues like facial expressions indicating stress. This oversight can lead to burnout and decreased morale within the team.

Research on emotional intelligence, such as by Goleman (1995), reveals that recognizing emotions—both our own and others'—requires not only awareness, but also attunement to non-verbal and contextual cues. Emotional blindness arises when individuals are distracted, stressed, or lacking practice in empathy, making emotional signals functionally "invisible." Thus, failing to "see" what's right in front of us emotionally can reflect ingrained habits of inattention or emotional distancing, not just deficits in visual or cognitive capacity, but in the ability to perceive the emotional landscape of our interactions..

Scenario: During a performance review, the manager is focused solely on metrics and fails to acknowledge the employee's emotional struggles. This leads to a detrimental work environment where employees feel unrecognized and unsupported.

Ideological Blindness (Social Psychology)

Ideological blindness refers to the inability to see beyond one's ideological beliefs, often resulting in a skewed perception of reality. This type of blindness can lead to confirmation bias, where individuals only perceive information that supports their pre-existing beliefs. For example, a partisan news consumer may disregard credible reports that contradict their beliefs about a political event, reinforcing their existing position.

Research by Hard (2014) and others details how deep-seated ideologies systematically narrow our mental field of view. These studies indicate that preexisting frameworks act as perceptual lenses, allowing in supporting information while screening out dissent. This means that even when contradictory facts or perspectives are in front of us, they can go unrecognized unless we deliberately interrogate our assumptions. "Not seeing" due to ideology is therefore a function of both cognition and motivational bias, shaping not just our interpretations but what we notice at all.

Scenario: In a discussion on national policy, individuals from opposing ideological backgrounds refuse to acknowledge data that challenges their views, leading to ineffective communication and increased division.

Inattentional Blindness (Neuroscience/Perception Studies)

Inattentional blindness is a psychological phenomenon where a person fails to perceive unexpected stimuli in their visual field when focusing on a specific task. This is due to the limited capacity of attention, which can cause significant oversights. For example, a driver focused on navigating a busy street may not notice a pedestrian stepping into the crosswalk, highlighting the dangers of inattentional blindness.

The classic work of Simons and Chabris (1999) and subsequent studies show that focused attention comes at a cost: Anything outside the scope of focus, no matter how obvious, may escape notice. This "blindness" is not about eyesight failure, but about the cognitive architecture that prioritizes tasks and filters background input. Thus, even when something is literally "right in front of me," the mind's spotlight can leave it in darkness—demonstrating how invisible the obvious becomes when we are absorbed elsewhere.

Scenario: During a distracted driving simulation, participants often fail to notice a cyclist, illustrating how focusing solely on the road can overlook crucial aspects of the environment. Increasing awareness of surroundings can improve overall safety.

The multiple forms of blindness enrich our understanding of the complexities of perception and awareness. By recognizing these dimensions, individuals can develop greater self-awareness and empathy, fostering healthier communication and decision-making in various facets of life. Understanding different types of blindness allows people to confront their limitations and strive for a more comprehensive understanding of their environments and interactions. We have a few more to consider.

Moral Blindness (Ethics, Philosophy)

Moral blindness refers to an inability or unwillingness to recognize ethical issues or moral implications of one's actions or beliefs. This blindness can lead individuals to act unethically without awareness or guilt, often because of a failure to critically evaluate their choices. For example, a corporate executive might prioritize profit over environmental sustainability, failing to see the broader moral implications of their business decisions.

Work in moral psychology and applied ethics, such as that by Peter Singer (1993), underscores that ethical perception relies on conscious reflection rather than automatic process. Researchers have found that organizational pressures, groupthink, and routine can desensitize individuals to the moral aspects of their own actions. As a result, “not seeing” the ethical implications of one’s choices is not generally about a lack of information, but rather about the moral numbing that comes from habituation and unexamined priorities—where wrongdoing is right there, yet functionally invisible to the actor.

Scenario: In a board meeting, a decision is made to cut costs by outsourcing to countries with inadequate labor laws, ignoring the ethical concerns about workers' rights and working conditions.

Perceptual Blindness (Cognitive Psychology/Neuroscience)

Perceptual blindness occurs when individuals fail to notice visual stimuli due to the limitations of their perception system. This blindness typically arises from cognitive overload, where the brain cannot process all the available information. For example, during a routine inspection, an office manager overlooks a significant malfunction in equipment because they are focused on other tasks, resulting in costly repairs and downtime.

Research following the work of Levin and Simons (1997) continues to show that our brains are wired to filter out what it deems nonessential in any given moment, ensuring that we are not overwhelmed by the sheer volume of sensory data. This selective attention, while adaptive, means that even significant elements directly in our line of sight can remain unnoticed if our cognitive resources are elsewhere. As a result, "not seeing what's right in front of me" often stems from our brain's active role in suppressing information outside our immediate focus, illustrating how perception is not just about visual input, but about what the mind prioritizes for conscious awareness.

Scenario: In a busy store, a shopper might be distracted by promotions and fail to notice a new arrangement of products that could benefit them, indicating how perceptual overload affects consumer behavior.

Political Blindness (Political Science)

Political blindness refers to a lack of awareness or acknowledgment of differing political opinions and ideologies. It can lead to polarization, where individuals or groups become entrenched in their beliefs without considering alternative viewpoints. For example, a political party exclusively engages with its supporters on social media, creating an echo chamber that reinforces its beliefs while remaining oblivious to the views of the opposition.

Sunstein's (2009) research and related studies on echo chambers show that social and media environments often reinforce our existing beliefs so strongly that dissenting information doesn't register, even when it is in plain sight. The group dynamics of polarization mean that "seeing" opposing views requires an intentional shift in perspective—an act that is regularly circumvented by curated information feeds. Thus, the inability to see what's politically obvious to others reveals not only a lack of exposure, but the powerful ways in which attention and belief are shaped by community, habit, and technological design.

Scenario: During a heated political debate, participants fail to consider opposing arguments, leading to a deadlock. If they had actively sought out and acknowledged differing opinions, they could have found common ground.

Psychological Blindness (Cognitive Psychology)

Psychological blindness refers to a cognitive bias that prevents individuals from perceiving or processing information accurately. This can involve overconfidence in one's beliefs or a failure to question existing assumptions. For example, a person may consistently overlook evidence that contradicts their beliefs about health and diet due to confirmation bias, leading to poor health choices.

Contemporary research into cognitive biases, including work by Kahneman (2011), finds that confirmation bias and related mechanisms shape not only what information we seek out, but what we're able to process at all. Evidence suggests that beliefs act as filters, blocking out information that doesn't fit our worldview—even when this information is as immediate as physical evidence or documented studies. In this way, "not seeing what's right in front of me" becomes a byproduct of our mind's efficiency at avoiding cognitive dissonance, causing us to overlook or dismiss inconvenient facts.

Scenario: An individual believes that a particular diet is universally effective and ignores research advocating for moderation or varied diets. Acknowledging diverse nutrition studies may enhance their health outcomes.

Social Blindness (Social Psychology)

Social blindness refers to an inability or difficulty in recognizing and understanding social cues, norms, and dynamics. This can include failing to pick up on body language, tone of voice, or context in social interactions. Social blindness can stem from various sources, such as neurodevelopmental conditions like autism spectrum disorder, or simply a lack of exposure to diverse social situations. It can lead to misunderstandings and misinterpretations in social engagements. For example, a person might not notice that a friend is upset because they miss subtle facial expressions or changes in tone that indicate distress.

Grandin and Barron (2005), both of whom are on the autism spectrum, share their personal experiences with social navigation. The book delves into why some individuals may find it challenging to perceive emotions and intentions that are apparent to others, serving as a guide to understanding the invisible barriers faced by those with social blindness and highlighting the importance of empathy and patience in fostering inclusive social environments.

Relative to our core focus, social blindness affects our ability to see what's right in front of us by impairing recognition and understanding of interpersonal dynamics, which influences personal and professional relationships. Individuals may overlook significant social interactions or fail to grasp the importance of nonverbal communication and emotional undercurrents in social settings. This can result in missed opportunities for connection, collaboration, and conflict resolution, as well as a lack of awareness of social changes that affect the broader community.

Scenario Example: Imagine a team leader who, due to social blindness, fails to notice the declining morale among team members. Despite signs of disengagement and tension during meetings, the leader remains focused solely on task completion and deadlines, ignoring the subtle signals indicating group dissatisfaction. This lack of awareness prevents them from addressing the issues promptly, leading to decreased productivity and a potential turnover. The leader's inability to perceive what's right in front of them in the form of social cues results in avoiding necessary interventions to boost team dynamics and morale.

Temporal Blindness (Behavioral Economics)

Temporal blindness is the failure to recognize the long-term consequences of one's choices, often leading to short-term thinking. This type of blindness can affect decision-making in areas such as health, finance, and environmental conservation. For example, a person might choose immediate gratification, like dining out frequently, while ignoring the long-term benefits of saving money for retirement or a significant purchase.

Behavioral economists following Thaler's work demonstrate that the future often fails to exert a strong pull on our present attention. Neurological studies show that our brains are wired for immediate reward, making future consequences abstract and easy to ignore, regardless of their eventual impact. "Not seeing" the future right in front of us—failing to anticipate the outcomes of current actions—is thus a product of both mental time discounting and the shortcomings of imagination, wherein delayed consequences do not register with the force of immediate experiences (Thaler and Larrick, 2000).

Scenario: During a financial planning session, an individual prioritizes spending on luxury items instead of investing in their future. Acknowledging the benefits of delayed gratification could have led to more prudent financial decisions.

Willful Blindness

Willful blindness is a legal and psychological concept where an individual intentionally chooses to remain unaware of facts or information that would require action or responsibility. In law, it pertains to a defendant's deliberate ignorance of illegal activities. In psychology and ethics, it relates to cognitive biases and moral responsibility in decision-making processes. It is often associated with ignoring inconvenient truths or avoiding accountability. For example, a company executive who chooses not to investigate allegations of fraud within their organization can be considered willfully blind.

Through case studies and research, Margaret Hefferman (2011) illustrates how individuals and groups avoid confronting uncomfortable truths, often leading to significant consequences. She dissects the psychological mechanisms and social drivers underpinning this behavior, such as confirmation bias and groupthink, and argues that acknowledging and addressing willful blindness is crucial for ethical decision-making and effective leadership.

Willful blindness directly pertains to the phenomenon of failing to see what is right in front of us by delving into the conscious and subconscious choices that lead people to ignore evident truths. This concept sheds light on how biases, societal pressures, and fear of repercussions can shape perception, leading individuals to actively discount or overlook critical information. In personal and professional contexts, this can result in missed opportunities, unethical actions, or harm that could have been prevented through proactive awareness and action. Understanding willful blindness helps to identify the psychological and environmental factors that contribute to selective attention and ignorance.

Scenario Example: Imagine a manager at a manufacturing plant where safety protocols are increasingly neglected, resulting in several near-miss accidents. Despite warning signs and employee concerns, the manager chooses not to see the escalating safety risks, focusing instead on meeting production targets and maintaining cost efficiency. Colleagues and the manager avoid acknowledging these dangers because addressing them might demand time-consuming changes and potential financial implications. This situation exemplifies willful blindness as a protective stance against uncomfortable realities. The manager's choice to remain blind to the issues right in front of them puts employees at risk and ultimately jeopardizes the company. This underscores how willful blindness can obscure obvious perils and underscores the importance of confronting difficult truths for the welfare of all involved.

Moving Through Blindness

Moving through various forms of "blindness" requires deliberate strategies that foster awareness, empathy, and critical thinking. Here's a structured approach that can help individuals and groups navigate through perceptual, emotional, cultural, psychological, political, change, moral, temporal, and ideological blindness.

Scenario Context: Let's build the scenario of a corporate team struggling with a lack of innovation due to ideological blindness. We include specific actions and situations that the team can take to move through their blindness. We also call out the transformative changes that are occurring with each pillar.

The following pillars represent the essential components of our approach. Rather than sequential steps, each pillar stands as a fundamental part of the overall framework, and addressing each one is necessary to achieve lasting success.

Pillar 1: Cultivating Awareness. *Self-Reflection:* Regularly engage in introspective practices such as journaling or meditation to recognize personal biases and emotions. *Feedback Loops:* Actively seek feedback from others about your perceptions and decisions. Open yourself to constructive criticism to identify blind spots. *Mindfulness:* Practice mindfulness techniques to enhance your awareness of both your thoughts and your environment. This can help mitigate inattentional blindness and improve focus. (Schwartz & Davidson, 2002)

Scenario Example: The team schedules a retreat for self-reflection. During this retreat, each member is encouraged to journal about their individual perceptions of the company's products and innovation processes. They complete a mindfulness exercise focusing on their biases in the workplace. At the end of the day, they share insights about what they might have overlooked regarding team dynamics and innovation.

Change: Beginning with self-reflection, team members become more aware of their biases and how those biases have stifled creativity

Pillar 2: Education and Exposure. *Diverse Experiences:* Engage with different cultures, ideologies, and perspectives through travel, community involvement, or educational programs. This can reduce cultural and ideological blindness. *Infinite Learning:* Stay informed about psychological, moral, and ethical issues. Participation in workshops, seminars, and lectures can help broaden your understanding of complex topics. (Nussbaum, 1997)

Scenario Example: The team leader organizes a field trip to a local startup known for its innovative culture. The team spends the day observing and interacting with the startup's employees, learning about their strategies for fostering creativity and collaboration. They also read articles and engage in discussions about emerging trends in their industry to broaden their perspectives and spark new ideas.

Change: Through exposure to new ideas and environments, they gain fresh insights that challenge their existing frameworks

Pillar 3: Critical Thinking. *Question Assumptions:* Regularly challenge your own beliefs and the beliefs of others. Use Socratic questioning to explore the foundations of your views. (See Appendix A.) *Analyze Sources:* When consuming information, critically assess sources, considering potential biases, motives, and perspectives that may not align with your own. (Paul & Elder, 2014)

Scenario Example: After returning from the field trip, the team holds a brainstorming session where members must present ideas for new products. However, before any idea can be accepted, each one must undergo a rigorous questioning session led by the team facilitator. Some ideas are challenged based on assumptions, and as a result, team members learn to articulate their viewpoints and refine their propositions.

Change: By employing critical thinking, they refine their product concepts and feel empowered to question assumptions.

Pillar 4: Encouraging Empathy. *Active Listening:* Practice active listening during conversations, demonstrating genuine interest in others' perspectives. This can improve emotional awareness and empathy. *Perspective-Taking:* Intentionally place yourself in others' situations to better understand their experiences and feelings. This can help mitigate moral and emotional blindness. (Goleman, 1995)

Scenario Example: The team conducts a role-playing exercise where each team member must take on the perspective of their primary customers. They create scenarios representing different customer experiences with their current products. After role-playing, team members debrief, discussing how these different perspectives can guide their future product development and innovation strategies.

Change: Encouraging empathy helps each team member understand their customer base more deeply, fostering a customer-oriented mindset

Pillar 5: Collaborative Approaches. *Diverse Teams:* Collaborate with a diverse group of people to benefit from varied insights. Whether in professional or social settings, diverse teams can help bring overlooked perspectives to light. *Structured Dialogue:* Use facilitation techniques in discussions where individuals may have opposing views. Ensure that everyone has an opportunity to voice their opinions without judgment. (Katzenbach & Smith, 1993)

Scenario Example: The team decides to implement a cross-functional project group that includes members from marketing, sales, and R&D. Each meeting involves presenting ideas and soliciting input from all members, ensuring that diverse voices are part of every discussion. This blend of perspectives prompts innovative ideas that wouldn't have emerged in traditionally siloed meetings.

Change: Collaborative approaches break down barriers between departments, leading to innovative solutions informed by diverse perspectives

Pillar 6: Goal Setting and Accountability. *Set Awareness Goals:* Create specific, measurable goals related to recognizing and addressing personal blind spots. For instance, aim to read a certain number of articles from opposing viewpoints each month. *Accountability Partners:* Work with others to hold yourselves accountable in addressing specific forms of blindness. Regular check-ins can help maintain focus and commitment.

Scenario Example: At the end of the retreat, the team sets specific goals for increasing innovation. They aim to present five new product ideas over the next quarter, each developed by mixed groups from different departments. Accountable partners are assigned, and they schedule bi-weekly check-ins to share progress and provide support, creating an environment that fosters collaboration and accountability. (Locke & Latham, 2002)

Change: Establishing goals and accountability keeps the momentum going and ensures that the team remains focused on their objective of enhancing innovation.

Pillar 7: Utilizing Technology. *Feedback Tools:* Use apps or platforms that allow for anonymous feedback from peers, which can help highlight areas of blindness of which you are not aware. *Learning Platforms:* Use apps or platforms that allow for anonymous feedback from peers, which can help highlight areas of blindness that you may not be aware of. (Chui, Manyika & Miremadi, 2016)

Scenario Example: The team adopts a collaborative platform, such as an idea management tool, where members can anonymously submit product ideas and feedback on others' suggestions. The platform also includes analytics that visually represents trends in submissions, helping the team understand which concepts resonate most and how they can align their efforts to exploit prevailing ideas.

Change: Finally, technology provides a platform for continuous improvement, enabling them to capture and evolve their ideas effectively.

Moving through various forms of blindness requires a multifaceted approach that combines self-awareness, education, empathy, collaboration, and critical thinking. Through this holistic approach, the corporate team not only overcomes their ideological blindness but also cultivates a culture of innovation and collaboration that significantly enhances their productivity and creativity. Each step reinforces the others, creating a synergistic effect that promotes growth and understanding within the team.

Chapter 10
Biases

Though answers may sit in our view,
Our *Whys* are too shallow, too few.
With blinders of mind,
We're biased and blind
Yet *Whysly* we choose what is true.

Let's talk about bias. The term *bias* has its roots in the Middle French word *biais* , itself derived from the Old Provençal *biais* , meaning "slant" or "oblique." Originally, it described a diagonal line or a slope, and by the 16th century in English, *bias* came to refer to a figurative slant—an inclination or predisposition for or against something. The term migrated from geometry to lawn bowling (where a "bias" is a weight that causes a ball to roll in a curved path), before taking on more abstract connotations related to judgment and perception.

By the 19th and 20th centuries, *bias* was increasingly used in psychology and the social sciences to describe systematic deviations from neutrality or objectivity. Early psychological investigations—such as those by Frederic Bartlett on memory and by later cognitive scientists—examined how biases emerged from the ways people processed information. Pioneers like Daniel Kahneman and Amos Tversky propelled the concept to the forefront of psychological research in the 1970s, showing that human decision-making is reliably shaped by cognitive shortcuts—"heuristics"—that produce consistent errors, or *cognitive biases*, in judgment. (A discussion of heuristics is included in Chapter 13, Bits and Pieces, under Cognitive Science.)

In its broadest sense, bias refers to a tendency or predisposition to think, act, or respond in certain ways, often unconsciously and outside rational control. Far from being a simple flaw, bias is deeply embedded in the operations of the human mind. Our brains evolved to process vast amounts of information quickly and efficiently, relying on shortcuts to make rapid judgments in uncertain or complex environments. While these shortcuts, or heuristics, enable survival, they also produce predictable errors in perception, reasoning, and self-understanding.

Biases are not solely a product of individual personalities or fixed psychological traits; rather, they are profoundly influenced by the specific situations and contexts in which people find themselves. While personal backgrounds, values, and experiences shape the types of biases one might be prone to, the environment—such as social settings, cultural norms, organizational structures, levels of stress, and even the presence or absence of certain information—can activate or suppress particular biases. For example, a person may be generally open-minded, yet succumb to confirmation bias when under time pressure or surrounded by like-minded colleagues. Similarly, situational factors like group dynamics, authority figures, or emotional intensity can sway even the most self-aware individuals. Therefore, recognizing that biases emerge at the intersection of the individual and the context underscores the importance of cultivating awareness not only of our own tendencies, but also of the environments and situations that might bring those tendencies to the forefront. This nuanced understanding is essential for effectively identifying and mitigating bias, both personally and within groups.

In the 21st century, the significance of bias has expanded in tandem with the explosion of digital information and the proliferation of misinformation (inaccurate or misleading information shared without intent to deceive) and disinformation (deliberate false or manipulated information intended to deceive). The sheer volume and speed of information, curated by algorithms that amplify existing preferences, make us more susceptible to our biases. Through selective exposure, echo chambers, and filter bubbles, individuals become more likely to encounter and accept information that confirms their biases, while discounting conflicting evidence. As a result, biases do not only affect personal judgments—they can distort entire social discourses and fuel polarization.

The contemporary discourse on bias is not just academic. Media organizations address unconscious bias in news reporting; technology companies train algorithms to recognize and mitigate bias in AI systems; workplaces seek to counteract implicit bias through diversity initiatives. Recognizing and addressing bias has become both a societal challenge and a personal imperative, critical for navigating a complex, information-rich, and often divided world. And there are many different types of biases that have been identified. Perhaps the best known is confirmation bias—which makes sense given the strong role of ego in the human—so, we will start there.

Confirmation Bias

Confirmation bias is a cognitive phenomenon where individuals tend to favor information that confirms their existing beliefs, opinions, or hypotheses while disregarding or minimizing evidence that contradicts those beliefs (Nickerson, 1998). "Preexisting beliefs or hypotheses" refer to the assumptions and conclusions individuals hold before encountering new information. These are established understandings or frameworks that guide how people interpret evidence. For example, someone might have a belief that a certain diet is the healthiest option and will only seek or remember information that supports that belief.

This bias can significantly influence various aspects of decision-making, information processing, and interpersonal interactions. Here are some key roles that confirmation bias plays:

Area	Role Played
Information Processing	Individuals are more likely to seek out, interpret, and remember information that is consistent with their preexisting beliefs. This selective exposure can shape their worldview and reinforce existing opinions. (Nickerson, 1998)
Decision-Making	In decision-making contexts, confirmation bias can lead individuals to focus on supportive evidence and ignore contrary information. This can contribute to poor judgments, especially in situations requiring critical evaluation of evidence, such as in business or healthcare decisions. (Tversky & Kahneman, 1974)
Social and Political Views	In social and political contexts, confirmation bias can contribute to polarization. Individuals often align themselves with groups or media that echo their beliefs, resulting in echo chambers and reduced exposure to diverse perspectives. (Gidron & Bonikowski, 2013)
Perception of Evidence	When confronted with ambiguous or incomplete information, people are likely to interpret it in a way that supports their existing views. This can lead to misinterpretations of data, events, or behaviors in everyday life, affecting relationships and societal discourse. (Plous, 1993).
Resistance to Change	Confirmation bias can make individuals resistant to changing their beliefs, even in the face of strong evidence to the contrary. This rigidity can hinder learning and personal growth, reinforcing existing stereotypes or misconceptions. (Lord, Ross & Lepper, 1979)
Impact on Scientific Inquiry	In scientific research, confirmation bias can affect the way hypotheses are tested. Researchers may unintentionally design studies or interpret data in ways that confirm their expectations while overlooking contradictory findings, potentially leading to flawed conclusions. (Ioannidis, 2005)

A common example of confirmation bias is found in beliefs about health and nutrition. An individual who believes that a particular diet is healthy may only seek out studies or testimonials that support this view while dismissing research that highlights potential risks or downsides.

Scenario Example: A student researching a topic might only seek out sources that support their thesis statement. They may dismiss well-reasoned arguments against their viewpoint, thus failing to develop a more nuanced understanding of the subject matter.

Confirmation bias falls primarily under the field of psychology, specifically, strongly within both cognitive psychology and social psychology. In cognitive psychology—which studies how people think, perceive, remember, and learn, confirmation bias is examined in the context of how individuals process information, form judgments, and make decisions based on their existing beliefs. In social psychology, the focus is on how individuals' thoughts, feelings, and behaviors are influenced by the social context and the presence of others. Researchers in social psychology explore how confirmation bias affects interpersonal relationships, group dynamics, and societal behaviors, particularly how individuals align themselves with like-minded groups.

Although not as prominent, confirmation bias is also studied in other fields. In behavioral economics the focus is on how cognitive biases, including confirmation bias, influences economic decision-making and consumer behavior. In political science, scholars analyze how confirmation bias shapes political beliefs and the polarization of opinions within societies. In philosophy, it is studied in the context of rational belief formation and epistemology (the study of knowledge). Overall, confirmation bias is a multifaceted phenomenon that spans various disciplines, reflecting its broad influence on human thought and behavior.

In terms of the focus of this book—as the tendency to search for, interpret, favor, and recall information in a way that confirms one's preexisting beliefs or hypotheses—confirmation bias leads individuals to overlook evidence that contradicts their beliefs. This makes it hard to accept, or even perceive, information that is essential for accurate recognition or understanding.

In addition to confirmation bias, other biases that relate to our core focus, which is why we can't see what is right in front of us, include the Availability Heuristic, the Anchoring Bias, the Cognitive Bias, the Dunning-Kruger Effect, the Implicit Bias, the Implausibility Bias, the Optimism Bias, and the Status Quo Bias. Each is briefly discussed below, includes an example and scenario, and is connected to our core focus.

Anchoring Bias

Anchoring bias occurs when individuals place undue weight on the first piece of information they encounter (the "anchor") when making decisions, affecting subsequent judgments. Initial information can have a lasting impact on subsequent judgments and decisions (Tversky & Kahneman, 1974). For example, in a salary negotiation, if an employer initially offers a low salary, that figure can anchor the candidate's expectations, making them less likely to ask for a significantly higher figure even if market research suggests they deserve more.

As in our example, anchoring biases occur because people often use the initial piece of information as a reference point in their decision-making process. The brain tends to latch onto the first piece of information it receives, regardless of its relevance. This cognitive shortcut simplifies decision-making but can lead to biased outcomes, especially when the anchor is arbitrary or flawed.

Relative to our core focus, anchoring bias helps explain why we often fail to see what is right in front of us: once an initial piece of information captures our attention and serves as an anchor, it distorts our perception of subsequent information—even if that information is more accurate or relevant. Our minds remain fixated on the anchor, making it difficult to objectively evaluate new evidence, facts, or perspectives that contradict the initial impression. This cognitive tendency means that individuals can overlook clear, obvious information simply because it conflicts with a pre-established anchor, causing them to ignore or undervalue what is actually present and relevant right in front of them. In essence, anchoring bias blinds us to more accurate or current information by locking our attention onto a reference point, making it difficult to truly see and assess what is plainly before us.

Scenario Example: A consumer browsing for a car sees an initial price tag of $30,000. When they later see a discounted price of $27,000, they may perceive the latter as a great deal, even if the fair market value of the car is closer to $25,000, due to the initial anchor.

Appointment Bias Drift

Appointment bias drift refers to the tendency for individuals or entities appointed to positions of influence to diverge from the expectations of their appointing authorities over time. Initially, these appointees may align closely with the goals or ideologies of those who selected them. However, as they gain experience in their roles and interact with evolving societal norms and pressures, their decisions and perspectives may shift. This drift highlights the unpredictability inherent in the appointment process, where factors such as legal precedents, public opinion, and personal values can lead to outcomes that do not necessarily conform to the original expectations of the appointing authority based on bias. Such dynamics can be observed across various sectors, including politics, corporate governance, and other fields, emphasizing the complexities of influence and the potential for independent judgment.

The Availability Heuristic

The availability heuristic is a mental shortcut that relies on immediate examples that come to mind when evaluating a specific topic, concept, method, or decision. Tversky and Kahneman (1973) introduce the concept of availability as a heuristic that influences people's judgments about the frequency of events based on how easily examples come to mind. For example, a person might perceive flying as more dangerous than driving because recent news stories have reported airplane accidents, even though statistically, flying is much safer than driving.

In relation to our focus, this heuristic can prevent individuals from accurately assessing the likelihood or importance of an event or situation. Because people often consider information that is most readily available or memorable, they may overlook the broader context or less visible evidence, leading to distorted perceptions of reality.

Scenario Example: A student preparing for an exam might recall only the topics they discussed with friends or those they reviewed recently, thus neglecting to study important but less memorable material that could appear on the exam.

Cognitive Bias

Cognitive bias refers to systematic patterns of deviation from norm or rationality in judgment, where our perceptions and decisions are influenced by subjective factors rather than objective reality, which distorts critical thinking. When discussing "subjective factors" in cognitive bias, this refers to personal experiences, emotions, beliefs, and motivations that shape an individual's interpretation of information. These factors can distort one's perception of reality, resulting in judgments that may not align with objective evidence or facts.

A study by Tversky and Kahneman (1974) showed that people tend to overweight small sample sizes when predicting probabilities, leading to overconfidence in their judgments. In relation to our focus, cognitive biases can distort our understanding and evaluation of information, making it difficult to acknowledge obvious truths or facts that contradict our preconceived notions.

Scenario Example: Imagine a manager who continues to believe that a particular marketing strategy is successful despite clear data showing falling sales. The manager disregards this evidence due to cognitive bias, focusing on past successes while ignoring current realities.

Dunning-Kruger Effect

The Dunning-Kruger effect refers to a cognitive bias where individuals with low ability at a task overestimate their competence, while those who are more skilled tend to underestimate their abilities. Research by Dunning and Kruger (1999) showed that a lack of skill can lead to an inflated self-assessment of competence, highlighting the disconnect between actual ability and self-perception. For example, a person who has just started learning to play an instrument may feel confident in their skills and believe they are ready to perform publicly, while more experienced musicians recognize the significant amount of practice required before reaching that level.

In relation to our focus, this bias can lead to a lack of self-awareness and an inability to recognize one's limitations, preventing individuals from identifying areas for improvement or seeking help. The overconfidence of the less skilled can blind them to reality, making it challenging to see the expertise that lies beyond their understanding.

Scenario Example: A new employee may believe they are highly proficient in project management after completing a simple task. They may then volunteer to lead a complex project without realizing the skills and experience required to manage it effectively, potentially leading to failure and negative consequences for the team.

Egocentric Bias

Egocentric bias is the tendency to overemphasize one's own perspective, contribution, or importance in situations involving others. For example, in a conversation, a person assumes the topic is as interesting or important to everyone else as it is to them, overlooking clear signs of disinterest or disengagement from others. Ross and Sicoly (1979) found that when individuals recall group events, they often attribute more responsibility to themselves than to others—a phenomenon known as the "availability of self-related information." This bias can distort perceptions of fairness, collaboration, and group dynamics.

Relative to our core question, egocentric bias can obscure obvious cues provided by others or in the social environment, as an individual's self-focus overshadows the reality of the broader situation. This prevents a clear view of collective contributions or needs that are visible but minimized in the mind's eye.

Scenario Example: Jessica, a project coordinator, recently led her team through a complex product launch. In her mind, Jessica recalls the many late nights she spent preparing reports, coordinating meetings, and double-checking logistics. When her manager asks about the challenges and successes of the project, Jessica emphasizes her own workload and the obstacles she personally overcame. As a result, she overlooks the fact that her teammate, Sam, repeatedly flagged confusion about certain product specifications in emails that Jessica only skimmed, assuming they weren't urgent. Likewise, Jessica forgets how another team member, Alisha, quietly managed client communications and resolved a major issue before it escalated. During the debrief, Jessica unintentionally credits herself with most of the success, barely mentioning her colleagues' contributions or the moments when they needed clarification. Blinded by egocentric bias, she misses clear cues—such as teammates' tentative questions or subtle expressions of fatigue—that the process wasn't as straightforward for everyone else. This skews her perception of the project and undermines genuine appreciation for the group's collective effort.

Implicit Bias

Implicit bias refers to the attitudes or stereotypes that affect our understanding, actions, and decisions in an unconscious manner (Banaji & Greenwald, 2013). These biases are often based on social identity. "Social identity" refers to the sense of self that individuals derive from their membership in social groups, such as nationality, ethnicity, religion, gender, or profession. It encompasses the values, norms, and beliefs associated with these groups and influences how individuals interact with others and perceive themselves in various contexts. For example, implicit bias has been demonstrated in hiring practices where similarly qualified candidates are evaluated differently based on their names, which may indicate their ethnicity.

Social identity theory, developed by Tajfel and Turner (1979), examines how individuals categorize themselves and others into groups (in-groups and out-groups), influencing behaviors, attitudes, and interactions. This field explores the dynamics of group membership, intergroup relations, and the impact on individual identity, particularly how social identities affect perceptions and behaviors towards others.

In relation to our focus, implicit biases can cloud our judgments and perceptions, leading us to make assumptions in social situations that obscure reality, especially when it comes to recognizing the capabilities or worth of individuals based on race, gender, or other identifiers.

Scenario Example: A teacher may unconsciously favor students who fit a certain stereotype (e.g., quiet, studious) over those who are more boisterous, thus overlooking the potential of a more outgoing student who could be high-achieving if given the chance.

Implausibility Bias

Implausibility bias refers to the tendency to dismiss information as improbable or unrealistic based on personal beliefs or experiences, often leading to overlooking or rejecting evidence that contradicts one's worldview (Leman, 2013). "Worldview" is the comprehensive perspective through which individuals perceive and interpret the world around them, shaped by factors such as culture, upbringing, education, and personal experiences. It includes beliefs about existence, morality, and human nature.

Each of the following examples of worldviews illustrates how deeply embedded beliefs and values can shape one's understanding and reactions to various aspects of life: (1) Religious Perspective—an individual who believes in a particular faith might view life events through the lens of that faith's teachings and tenets; (2) Secular Humanism—someone who subscribes to secular humanist beliefs may place emphasis on reason, ethics, and justice while viewing the world through a scientific and rational lens; (3) Environmentalism—a person with a strong environmental worldview may prioritize ecological sustainability and perceive industrial practices as harmful, impacting their opinions and actions regarding development and policy; and (4) Cultural Framework—a person with a strong environmental worldview may prioritize ecological sustainability and perceive industrial practices as harmful, impacting their opinions and actions regarding development and policy.

In relationship to our focus, implausibility bias can cause individuals to fail to accept facts or alternative perspectives that seem implausible, preventing them from recognizing the truth before them. For example, wide-spread disbelief in climate change despite overwhelming scientific evidence. Many individuals find the implications of climate change too implausible based on their lived experiences.

Scenario Example: A team working on a technology project might dismiss an innovative idea from a junior member. They believe that it is too far-fetched or implausible, failing to see its potential value just because it deviates from traditional methods.

Ingroup/Outgroup Bias

Ingroup/outgroup bias is a social psychological phenomenon where individuals exhibit preferential treatment and positive attitudes toward members of their own group (the ingroup) while displaying prejudice or negative attitudes toward those outside their group (the outgroup). This bias can shape perceptions and interactions, influencing how people see themselves and others. For example, a sports fan might believe their favorite team's players are more talented and honest than those of rival teams, regardless of actual performance statistics.

Henri Tajfel's (1982) work on social identity theory examines how individuals derive a sense of identity and self-esteem from their group memberships. This research demonstrates how categorizing people into groups leads to ingroup favoritism and outgroup discrimination, even in minimal group situations where group distinctions are arbitrary.

Relative to our focus, ingroup/outgroup bias affects our ability to see what's right in front of us by distorting perceptions of reality based on group affiliations. This bias can lead individuals to overlook negative traits or behaviors within their own group while exaggerating flaws in outgroup members. Such distorted perceptions can prevent people from recognizing common ground or valuable perspectives from outside their group. Consequently, this bias reinforces divisive thinking and impedes open-minded interactions with differing groups, affecting personal, professional, and societal dynamics.

Scenario Example: Imagine an office environment where two departments, Marketing and Sales, experience tension due to ingroup/outgroup bias. Members of the Marketing team tend to attribute their own success to hard work and intelligence, while viewing the successes of the Sales team as merely luck or manipulation. This bias blindsided them to the positive contributions of their colleagues in Sales, creating unnecessary conflict and undermining collaboration. By failing to see the strengths and efforts of the other team right in front of them, both groups miss opportunities for synergy and innovation that could arise from a more unified approach.

Negativity Bias

Negativity bias is a cognitive bias whereby negative events, emotions, or information exert a greater impact on psychological state and processes than equally intense positive ones. Baumeister et al. (2001) describe this as the principle that "bad is stronger than good," documenting how individuals are more likely to focus on, remember, and be influenced by negative stimuli in their environment. Early in human history, this evolutionary adaptation may have offered survival benefits, as being alert to potential threats was critical, but in modern context it can distort perception and decision-making. For example, an employee receives largely positive feedback but fixates on a single minor criticism, overlooking overall success.

Baumeister et al. (2001) provide a comprehensive review of research on negativity bias across multiple domains—including learning, emotion, social interactions, and memory—clearly establishing that negative events and emotions have a more powerful and lasting impact than positive or neutral events. Their work underscores the pervasive influence of this bias in shaping human perception and behavior.

In the context of our core focus, negativity bias can prevent individuals from seeing opportunities, strengths, or positive developments that are present but less emotionally salient. When people are predisposed to attend to what is wrong or threatening, they may overlook the broader picture or obvious positives directly in front of them, resulting in a skewed perception that emphasizes risk or loss at the expense of benefit or progress.

Scenario Example: A project manager receives overwhelmingly positive feedback about a recent launch, but becomes preoccupied with a single negative comment from a client, failing to recognize the overall success that is clearly supported by the majority of evidence.

Normalcy Bias

Normalcy bias is a cognitive bias that causes individuals to underestimate both the possibility of a disaster and its potential impacts, assuming that things will continue as they always have. Omer and Alon (1994) discuss how this bias leads people to dismiss warnings and delay taking precautionary action, even in the face of mounting evidence of impending crisis. This bias can foster inaction and denial, particularly during emergencies, and is often observed in situations ranging from natural disasters to public health crises. For example, residents ignore evacuation orders before a hurricane, believing the situation "can't get that bad here."

As it relates to our overall focus, normalcy bias limits our ability to see and respond to risks or changes that are clearly present or imminent. Because the mind instinctively favors the predictability of current routines, individuals may remain blind to evidence indicating that action or preparation is needed, simply because such events are unprecedented in their experience.

Scenario Example: A major technology company receives early warning signs of a cyberattack targeting its outdated security systems. Several employees notice unusual network activity and report it to their supervisors, but the leadership team dismisses these concerns, believing the company's systems have always been secure. They assume such a breach could never happen to them because it never has. As a result, they fail to take timely action or upgrade their security protocols, overlooking the clear signs of risk right in front of them. When the attack finally occurs, the company faces significant data loss and reputational damage—damage that could have been prevented had the warnings been acknowledged and acted upon.

Omission Bias

Omission bias is a cognitive bias that occurs when individuals judge harmful actions as worse, or less moral, than equally harmful omissions (inactions). In other words, people tend to favor not doing something that leads to a negative outcome rather than taking actions that might lead to the same outcome. This bias can lead to poor decision-making in areas like healthcare, law, and environmental policy. For example, consider a doctor faced with two treatment options for a patient with a terminal illness: one treatment has a known risk of severe side effects while the other involves not treating the patient at all. The doctor might be more inclined to choose the option of doing nothing, viewing it as morally preferable and less likely to cause harm, even though not treating the patient could lead to significant suffering.

Millar and Millar (2021) explore the concept of omission bias and its implications for ethical decision-making in organizations, illustrating how individuals often exhibit a preference for inaction over actions that may result in similar negative consequences. Through a series of experiments, they found that participants were less likely to support actions that might harm an individual, even when the omission of an action would lead to greater overall harm. This research highlights how omission bias can skew moral reasoning and lead to suboptimal outcomes in both personal and corporate settings, emphasizing the need for awareness of cognitive biases in everyday decision-making.

Relative to our core focus, omission bias can contribute to the phenomenon of failing to see what's right in front of us because it often fosters a mindset of avoidance. When individuals focus on the potential negative outcomes of taking action, they tend to ignore or downplay the negative consequences of inaction. This can lead to a skewed perception of reality where the more harmful implications of doing nothing are overlooked, resulting in a difficulty in recognizing the solutions that are visible yet may require assertive action. This cognitive dissonance creates a barrier to effective decision-making, as individuals grapple with the discomfort of taking responsibility for actions that might lead to perceived harm.

Scenario Example: Imagine a school board deciding whether to adopt a new digital learning program. Some board members feel that implementing the program has risks – it could lead to less personal interaction among students or technical issues. Meanwhile, a decision to maintain the current system could result in lagging student engagement and poor learning outcomes. Omission bias might lead the board to prefer the "do nothing" option of sticking with the current system despite clear evidence of its inefficacy, because they perceive the risk of acting (implementing the program) as more severe than the risks of inaction.

Optimism Bias

Optimism bias is the tendency to believe that one is less likely to experience negative events compared to others. Research by Weinstein (1980) explores the concept of unrealistic optimism, demonstrating how some individuals perceive themselves as less vulnerable to negative events than others. For example, smokers often believe they are less likely to develop health problems than the general population, despite the well-documented risks associated with smoking.

Relative to our core focus, this bias can lead individuals to underestimate risks and ignore potential setbacks or hazards. As a result, they may miss vital signs or warnings about issues because they assume they themselves will not be affected.

Scenario Example: A student may underestimate the difficulty of a course and believe they can pass the final exam without studying adequately, even after observing peers struggle with the material.

Projection Bias

Projection bias refers to the tendency to assume that others share one's own beliefs, values, emotions, or current mental states, while also failing to anticipate changes in one's own preferences over time. For example, a manager assumes all employees are motivated by public recognition, missing signs that some value privacy, thus overlooking obvious cues. Loewenstein, O'Donoghue, and Rabin (2003) argue that projection bias leads to systematic errors in predicting others' behavior and in making decisions for one's future self, as people "project" their own immediate feelings and perspectives onto others or into the future.

In the context of not seeing what's plainly before us, projection bias blinds us to the true diversity of others' perspectives, needs, or motivations. By assuming everyone thinks or feels as we do, we can overlook clear cues or evidence that they do not, impeding effective communication and mutual understanding.

Scenario Example: Imagine a team leader, Jonas, who thrives on fast-paced, high-pressure deadlines and feels motivated by public praise in meetings. When planning a challenging new project, Jonas sets an aggressive timeline and announces a "Team Member of the Month" award, convinced that these incentives will drive everyone's best performance—because this is what would work for Jonas.

However, several team members are actually more creative and productive with quieter, flexible work environments. One team member, Jamie, feels anxious about public recognition and would prefer private feedback. Another, Pat, juggles family obligations and values a more balanced schedule. Jonas, blinded by projection bias, fails to notice when Jamie grows withdrawn after being put on the spot in a meeting, or when Pat seems stressed and expresses concerns about extra hours. Despite these clear signals, Jonas interprets any lack of enthusiasm or stress as a lack of dedication, not realizing that the team's needs and motivations differ from his own. As a result, communication falters, morale dips, and project outcomes suffer—all because Jonas projected his own preferences onto the team and overlooked the evident signs that others had different, plainly observable needs.

Self-Serving Bias

Self-serving bias is the tendency for individuals to attribute their successes to internal, personal factors (such as talent or effort), while attributing failures to external circumstances. This automatic pattern of attribution serves to protect self-esteem but can prevent objective self-assessment. For example, a student blames a poor exam score on an "unfair test" but attributes good scores to their intelligence, missing the need to address their study habits.

Miller and Ross (1975) highlight how this bias is evident in a wide array of judgments, from academic performance to workplace evaluations, and suggest that self-serving attributions function to maintain a positive self-image in the face of failure or criticism.

With respect to our focus, self-serving bias can obstruct recognition of clear truths about one's performance, limitations, or contributions. By favoring interpretations that protect self-image, individuals may ignore evidence of mistakes, overlook areas in need of improvement, or misattribute the causes of outcomes, remaining blind to facts that are objectively apparent.

Scenario Example: An employee consistently claims credit for team successes by emphasizing their leadership, but when projects fail, blames external factors like poor market conditions, thereby missing opportunities for self-improvement that are directly observable.

Status Quo Bias

Status quo bias is the tendency to prefer the current state of affairs and resist change, even when alternatives may be more beneficial. For example, a company may continue using outdated software due to the discomfort of switching to a new system, despite evidence that the new software would enhance productivity.

There are specific indicators that cause individuals to favor the current situation over alternatives, providing insights into the psychological mechanisms behind status quo bias (Samuelson & Zeckhauser, 1988). One key indicator is *loss aversion*—the tendency to perceive potential losses from change as more significant than the potential gains. People often weigh the risks of leaving the status quo more heavily than the rewards of adopting something new, which makes them overly cautious. Another indicator is the *exposure effect:* individuals naturally develop a preference for things simply because they are familiar. This familiarity is comforting and can subconsciously reinforce the belief that the current state is preferable, regardless of evidence.

A third indicator is the *sunk cost fallacy*. This refers to the tendency to continue investing resources (time, money, effort) into a current situation because of the resources already committed, even when it may no longer be the best option. Individuals may choose to stay with an outdated software system simply because of the time and money already spent on it, rather than considering the long-term benefits of switching to a more efficient alternative. A fourth indicator is the *fear of uncertainty* related to change. The discomfort associated with the unknown can prompt individuals to cling to the familiar, even at the expense of improvement. This fear of uncertainty can lead to avoidance of new technologies, practices, or strategies despite clear evidence of their advantages. This bias is also tied directly to cognitive dissonance and the anchoring effect described above.

Relative to our focus, this This bias can lead individuals to overlook better options or solutions, resulting in a reluctance to innovate or adapt. It maintains the status quo and can prevent the recognition of needed changes that are clear to outside observers.

Scenario Example: A neighborhood might resist the construction of a new park, preferring to maintain the current landscaping despite clear environmental benefits and community support for the development.

Each of these biases plays a crucial role in shaping how we perceive the information around us and often leads to an inability or unwillingness to "see" the reality that contradicts our beliefs or experiences. By becoming aware of these biases, we can strive to mitigate their impact on our understanding and decision-making processes.

Learning to Recognize Bias

In the previous chapter's scenario example of a corporate team working to overcome ideological blindness and cultivate innovation (titled "Moving Through Blindness"), several biases became apparent at different pillars of the process. *Reflect:* Did you recognize any of these? Were you uncomfortable with the scenario example? To hone your skill at recognizing bias, go back and take another read. Then, below, is a breakdown of the specific biases emerging during each relevant step along with an explanation.

Pillar 1: Cultivating Awareness

Biases Identified: Self-serving bias and confirmation bias.

Explanation: As team members engaged in self-reflection, they became aware of how they might have favored information or ideas that supported their previous success (self-serving bias) and overlooked contrary perspectives or evidence (confirmation bias). This step helped them recognize that their past experiences and successes might have clouded their judgment regarding innovation.

Pillar 2: Education and Exposure

Biases Identified: Groupthink and cultural bias.

Explanation: Exposure to the local startup revealed how groupthink had led the team to arrive at conclusions without critically assessing alternative ideas. They also recognized cultural bias, understanding that their preconceived notions about best practices in their industry might limit their creativity and problem-solving abilities.

Pillar 3: Critical Thinking

Biases Identified: Anchoring bias and availability heuristic.

Explanation: Exposure to the local startup revealed how groupthink had led the team to arrive at conclusions without critically assessing alternative ideas. They also recognized cultural bias, understanding that their preconceived notions about best practices in their industry might limit their creativity and problem-solving abilities.

Pillar 4: Encouraging Empathy

Biases Identified: Empathy gap and perspective bias.

Explanation: The role-playing exercise helped team members identify their empathy gap—the disparity between understanding others' emotions and genuinely feeling them. Additionally, they recognized perspective bias, realizing that they had primarily viewed issues from their own viewpoints without considering customer needs holistically.

Pillar 5: Collaborative Approaches

Biases Identified: In-group bias and stereotyping.

Explanation: As the team engaged in cross-functional collaboration, they became aware of in-group bias—favoring opinions and ideas from their own departments over others—and how stereotyping had informed their early perceptions of colleagues from other departments, which limited innovation.

Pillar 6: Goal Setting and Accountability

Biases Identified: Planning fallacy and optimism bias.

Explanation: While setting ambitious goals, team members began to recognize the planning fallacy, where they underestimated the time and resources needed to complete tasks. The optimism bias also surfaced, with individuals tending to overestimate the likelihood of positive outcomes without weighing potential challenges realistically.

Pillar 7: Utilizing Technology

Biases Identified: Confirmation bias (again) and technological determinism.

Explanation: As they used the idea management tool, some team members had to confront their tendency to favor ideas that aligned with their previous beliefs (confirmation bias). They also noticed a creeping form of technological determinism—believing that the technology alone would drive innovation without considering the need for active participation and creativity from the team.

Throughout the various steps, the team became aware of multiple biases, including self-serving bias, confirmation bias, groupthink, cultural bias, anticipation and planning fallacies, empathy gaps, and others. Recognizing these biases allowed them to address their ideological blindness and foster a more innovative and collaborative environment, ultimately leading to improved decision-making and creativity within the organization

Identifying Your Own Biases

There's no single checklist to uncover every bias, as they are numerous and often dependent on situation, context, and an individual's own psychology and experiences. However, a reflective, metacognitive approach can be very useful. Here are some reflective questions commonly recommended by psychologists, ethicists, and behavioral scientists to help individuals identify and challenge their own biases, regardless of the specific type.

1. *What assumptions am I making right now?* Are these assumptions based on evidence, or are they shaped by my prior experiences or beliefs?

2. *Am I considering multiple perspectives, or just my own?* Whose voices, needs, or viewpoints might I be overlooking or dismissing?

3. *How might my background, culture, or experiences influence how I interpret this situation?* Could someone with a different background see this differently?

4. *What evidence exists that challenges my perspective or decision?* Do I automatically dismiss information that doesn't fit how I already see things?

5. *Am I confusing my personal preferences, values, or emotions with objective reality?* Do I automatically dismiss information that doesn't fit how I already see things?

6. *Have I sought out feedback or input from people with different experiences than mine?* How open am I to criticism or alternative viewpoints?

7. *When was the last time I changed my mind about something meaningful?* What led me to do so? Am I too anchored in previous opinions?

8. *If someone critiqued my thinking, would I feel defensive, or curious?* What does my reaction reveal about my openness to recognizing bias?

9. *In what situations do I tend to make snap judgments?* How often do I pause to reflect, and what happens when I do?

10. *How will I feel about this decision/attitude in a week, month, or year?* Am I projecting my current feelings or preferences onto my future self or others?

These questions are not exhaustive, but they offer a flexible framework you can revisit in different circumstances and with various types of bias (projection bias, confirmation bias, groupthink, etc.). *Tip:* Making a habit of pausing and asking these questions—especially when the stakes are high or when you notice strong emotional reactions—can slowly help uncover and weaken the grip of hidden biases.

After Thought

To understand bias is to recognize that our perceptions of reality are filtered through layers of cognitive, emotional, and social inclinations—many of which operate outside our conscious awareness. In an age marked by both unprecedented access to information and the hazards of misinformation and disinformation, cultivating awareness of our own biases is essential. Only by doing so can we hope to "see what is right in front of us"—and to engage with the world more accurately, empathetically, and thoughtfully.

Chapter 11
Cognitive Dissonance

In dissonance our thoughts often spin,
As clashing beliefs make a din.
We *Whysly* explore,
To balance and more,
Finding harmony hidden within.

Cognitive dissonance is the psychological phenomenon that occurs when an individual holds two or more contradictory beliefs, values, or attitudes, leading to feelings of discomfort and tension. This discomfort often motivates individuals to reduce the dissonance either by reassessing their beliefs, acquiring new information, or changing their behaviors.

Cognitive dissonance plays a crucial role in our inability to "see" what is right in front of us. When faced with information that contradicts pre-existing beliefs or attitudes, people experience psychological discomfort. Rather than acknowledging conflicting evidence or adjusting their views accordingly, individuals may opt for rationalization, denial, or selective perception to maintain consistency. This can result in a blindness to realities that strongly conflict with their established belief systems.

While certainly understandable, what is fascinating to explore are the similarities and differences as researchers across a myriad of fields and disciplines look at cognitive dissonance through their unique viewpoints. Taking a consilience approach and reflecting on these varied perceptions, there is much to learn about both the human mind as well as cognitive dissonance. Each field provides a unique definition and perspective on cognitive dissonance, allowing us to deepen our understanding of how this phenomenon manifests in different contexts. Let's take the time to become consilient explorers, and take a closer look at how cognitive dissonance is perceived from the viewpoint of education, marketing, political science, psychology, and sociology, and how these differences can inform our understanding regarding why we struggle to "see" what is in front of us.

Education Viewpoint

In education, cognitive dissonance is defined as the conflict students experience when their existing beliefs or knowledge are challenged by new information or experiences that contradict their prior understanding. This definition underscores the role dissonance plays in learning. It can serve as a motivator for students to engage with material that conflicts with their previous beliefs, ultimately encouraging critical thinking and personal growth.

Research by Greene and Paxton (2009) applies cognitive dissonance theory to educational contexts, exploring how dissonance can impact moral and ethical decision-making among students. The study demonstrates how awareness of one's inconsistency can lead to changes in both attitude and behavior in learning environments. For example, a student who cheats on an exam may initially feel justified due to pressure but ultimately experiences dissonance, leading them to change their views on integrity.

Relative to our focus on why we can't see what is right in front of us, understanding cognitive dissonance in educational settings emphasizes the importance of fostering an environment where challenging existing views is welcomed. This can help students "see" complexities in their understanding and better embrace new, potentially contradictory evidence.

First Scenario Example: In a high school history class, students are learning about the causes of World War II. Many students enter the class with the belief that economic factors alone—such as the Great Depression—were the primary cause of the war. As the teacher presents new information about the rise of totalitarian regimes, including ideological and political factors, the students begin to experience cognitive dissonance.

One student, Sarah, has always believed that her understanding was complete, and she struggles to reconcile this new information that emphasizes the power of propaganda and nationalism in countries like Germany and Italy. Her existing belief system is challenged when the teacher shares video clips from that era illustrating how leaders manipulated public opinion.

As the lessons progress, Sarah finds herself increasingly uncomfortable with this contradiction. She reflects on her previously held beliefs and, after discussions with classmates about the complexities of the political landscape of the time, she begins to see that multiple factors contributed to the war, not just economic ones. This leads her to reevaluate her understanding of historical events and to engage more deeply with the material by researching how various nations' ideologies influenced their actions.

Ultimately, this scenario highlights how cognitive dissonance can stimulate critical thinking and encourage students like Sarah to expand their viewpoints and develop a more nuanced understanding of history. Let's create a second scenario that relates to the current U.S. government cuts of funding and people related to education programs.

Second Scenario Example: In a public university, a group of students is engaged in a political science class discussing the role of government in education. Many students believe that education is a right for every person who chooses to improve themselves and their contribution to society. Other students have been taught that government efficiency leads to better resource allocation and improved educational outcomes, but this belief is challenged when they learn about recent governmental actions aimed at cutting funding for educational programs, and even attempts to eliminate the Department of Education.

As the professor presents case studies of universities currently facing significant budget cuts, mass layoffs of faculty and support staff, elimination of scientific and medical research, and barriers put in place for acceptance of foreign faculty and students, significant cognitive dissonance is experienced. For example, one student, Aram, has previously believed that reducing government spending is a universally positive approach to improving efficiency. However, as he sees how these funding cuts directly impact their university—leading to loss of diverse thinking, reduction of available expertise, larger class sizes, reduced course offerings, and lower support for vulnerable students—he begins to question the validity of his previous beliefs. He also begins to reflect that efficiency in government—and in the private sector as well—cannot be implemented at the cost of effectiveness.

In a class debate, some students argue that government interventions can sometimes be detrimental, and others, echoing Aram's reflection, insist that efficiency should not come at the expense of education quality. Aram finds himself wrestling with conflicting ideas: How can efficiency, which sounds beneficial in theory, lead to negative outcomes in practice? This dissonance motivates Aram to refocus on quality and research alternative viewpoints on education funding, reflect on the power of diversity, explore the importance of maintaining strong educational support systems, and engage in conversations about the ethical implications of current government policies.

Ultimately, this scenario highlights how cognitive dissonance can provoke critical thinking and deeper engagement with complex issues, encouraging students to reassess their beliefs about government actions in education and the true meaning of efficiency in this context.

Marketing Viewpoint

In marketing, cognitive dissonance refers to the *discomfort consumers feel when their purchase decisions conflict with their beliefs* about a product's value or the relevance of the purchase to their self-image. This perspective highlights how dissonance can shape consumer behavior and decision-making. Marketers often use strategies to create dissonance that encourages individuals to justify their purchases, reinforcing brand loyalty and mitigating potential regret. For example, a consumer who buys an expensive product might convince themselves of its necessity by discounting less costly alternatives, thereby easing their dissonance.

Festinger and Maccoby (1964) highlight how advertising can create dissonance and subsequently influence buyer decisions through persuasive techniques promoted by advertisers such as presenting testimonials that align with the target audience's values. Additionally, advertisers often employ *scarcity tactics*, suggesting that a product is in limited supply, which can instill a sense of urgency and consequently reduce cognitive dissonance associated with the purchase. Another technique is the use of *emotional appeals*, where advertisements evoke feelings of happiness, nostalgia, or fear to connect with consumers on a personal level, thereby reinforcing the perceived value of the product. Lastly, the strategy of *framing a product in a positive light* by emphasizing its unique benefits or advantages over competitors can help shift consumer focus away from any doubts they may have, further mitigating dissonance and promoting a favorable perception. These techniques collectively work to influence consumer behavior and solidify purchase decisions, despite lingering uncertainties.

In the realm of customer behavior, when individuals make a purchase and later experience uncertainty about it, they may employ various persuasive techniques to alleviate their discomfort. One common method is the use of *commitment bias* (see Chapter 10), where consumers emphasize the effort they expended in making the choice, thereby justifying their purchase. Another technique is *social proof*, where individuals seek validation from others who share their views or have made similar purchases, reinforcing their choice and diminishing doubt. Lastly, the *anchoring effect* can come into play; this is when consumers fixate on the initial price or perceived value of a product, leading them to disregard alternatives that might offer better benefits. These strategies contribute to a distorted perception of the product and its value, creating a narrow view that resists reevaluation.

Relative to our core focus, insights from marketing illustrate that people may rationalize purchases they later feel uncertain about, blinding them to alternatives that may be better suited. This inherent dissonance can distort their perception of the product and its value, creating a narrow view that resists reevaluation.

Scenario Example: Emma just bought an expensive phone after weeks of research. Once she overhears a friend's negative review about it, she feels dissonance. To reduce this discomfort, she posts positive experiences on social media, reinforcing her choice publicly.

Political Science Viewpoint

From a political science perspective, cognitive dissonance is viewed as the tension that arises when an individual's political beliefs or party affiliations are challenged by new information or behaviors from their favored party. This definition points to how cognitive dissonance can reinforce political polarization. When confronted with information that contradicts their political identity, individuals may resort to denial or rationalization, further entrenching their biases rather than allowing for constructive discussions. For example, a voter may support a candidate whose policies conflict with their core values, leading to dissonance that they resolve by focusing on the candidate's qualities rather than the policies.

Zeller's (1992) work provides insight into how cognitive dissonance shapes public opinion and political behavior. It illustrates how individuals adjust their attitudes to align more closely with public discourse, leading to a malleable understanding of political realities influenced by social pressures and cognitive dissonance.

In relation to our focus, this perspective helps explain why individuals often cannot "see" opposing political viewpoints. The discomfort associated with encountering contradictory information can lead to an increased commitment to previously held beliefs, preventing productive dialogue and mutual understanding.

Scenario Example: Lisa supports a political candidate due to their charisma but feels conflicted when learning about policies that contradict her values. To ease this dissonance, she begins to cite reasons for supporting the candidate that align with her values, thereby convincing herself that the candidate will change their policies once in office.

Psychology Viewpoint

In psychology, cognitive dissonance is defined as *the mental discomfort experienced when an individual holds two or more contradictory beliefs, values, or attitudes simultaneously* or when their *behaviors conflict with their beliefs*. This psychological framework emphasizes internal conflict and self-identity. It helps explain why individuals may ignore or dismiss contradictory evidence that threatens their self-concept. The discomfort might lead one to modify attitudes or rationalize behaviors to achieve harmony.

In relation to our core focus, Understanding cognitive dissonance in psychology reveals that discomfort can lead to avoidance behaviors. When faced with conflicting beliefs, rather than acknowledging inconvenient truths, individuals often cling to their existing beliefs, blinding them to realities that may serve their interests better. For example, a smoker may continue smoking despite knowing the health risks because they rationalize that their grandfather smoked and lived a long life.

In his groundbreaking work outlining the theory of cognitive dissonance, Festinger (1957) describes the conditions under which dissonance arises and the various ways individuals cope with it. Festinger explains how dissonance can influence attitudes, behaviors, and decision-making processes, forming a cornerstone of modern psychological theory. This work fundamentally reshaped how psychologists understand the connection between internal beliefs and outward behavior. According to Festinger, when individuals are faced with situations that produce dissonance—such as holding two contradictory beliefs or acting in a way that conflicts with their self-image—they experience psychological discomfort. This discomfort acts as a powerful motivator, prompting individuals to resolve the dissonance and restore internal harmony. Often, this resolution manifests through changes in attitudes or beliefs to more closely align with one's actions, or by justifying behaviors to reduce the sense of conflict. For example, someone who makes a difficult decision may later emphasize the positive aspects of their choice while downplaying any drawbacks, thus reshaping their attitude to fit their behavior.

This dynamic has significant implications for decision-making processes. The desire to avoid or reduce cognitive dissonance can lead people to ignore, dismiss, or reinterpret information that challenges their established beliefs or values, ultimately influencing the choices they make. Festinger's insights reveal that rather than systematically evaluating evidence, individuals may unconsciously shield themselves from uncomfortable truths, leading to biased judgments and resistance to change. By highlighting these mechanisms, Festinger's work not only elucidates why people might rationalize harmful or contradictory behaviors—such as a smoker dismissing health warnings—but also underscores the powerful influence of internal conflict on both everyday decisions and long-term attitudes.

Scenario Example: Sarah is a health-conscious individual who has always criticized fast food. However, one evening, feeling rushed, she decides to grab a burger. Instead of feeling guilty, she tells herself that it's fine to indulge occasionally, creating a justification to alleviate her dissonance.

Sociology Viewpoint

From a sociological perspective, cognitive dissonance is understood as *the conflict that arises when individuals' beliefs are challenged by social norms or group dynamics*. This can lead to a stronger adherence to group beliefs in the attempt to maintain social identity. For example, A group of friends may adopt certain opinions about a political candidate because that is the prevailing view among the group, then rationalize their support despite conflicting evidence.

This focus on social identity highlights how external pressures influence individual beliefs, showing that dissonance can be aggravated by the need for acceptance within a group. Individuals may adjust their personal beliefs to align with the group to avoid social ostracism, often ignoring personal evidence to maintain group cohesion.

The research by Turner and Hogg (1987) suggests that dissonance can arise not just from personal beliefs but from group norms and the pressure to confirm. This pressure often compels individuals to align their views with those of their social group, as doing so strengthens their sense of belonging and adherence to group identity. Individuals may internalize group perspectives to such an extent that they prioritize group cohesion over objective analysis, reinforcing a cycle where dissent is discouraged and conformity is celebrated.

In relation to our core focus, this sociological angle highlights that the desire to fit in with one's social circle can significantly inhibit critical thinking and the acceptance of conflicting information. Individuals may become so entrenched in their group's beliefs that they actively avoid engaging with evidence that contradicts their views, prioritizing acceptance within the group over the pursuit of truth. This dynamic illustrates how social identity can corrupt rational decision-making, create blind spots and blinks, and create barriers to open discourse, ultimately perpetuating misinformation and ideological uniformity.

Scenario Example: Mark is part of a social group that shares anti-vaccine views. When presented with data supporting vaccinations, he feels dissonance. Instead of considering the evidence, he dismisses it as part of a "mainstream conspiracy," reinforcing his group's narrative to ease the discomfort.

Comparison and Contrast

Let's use a table to look at how the contrast across these four fields, focusing on definitions, understanding, and relationship to our core focus.

Field	Definition	Understanding	Relationship to Focus
Psychology	The mental discomfort experienced when an individual holds two or more contradictory beliefs, values, or attitudes simultaneously or when their behaviors conflict with their beliefs.	Emphasizes internal conflict and self-identity. It helps explain why individuals may ignore or dismiss contradictory evidence that threatens their self-concept.	Reveals that discomfort can lead to avoidance behaviors. When faced with conflicting beliefs, rather than acknowledging inconvenient truths.
Sociology	The conflict that arises when individuals' beliefs are challenged by social norms or group dynamics.	Highlights how external pressures influence individual beliefs, showing that dissonance can be aggravated by the need for acceptance within a group.	Explain why social context can inhibit one's ability to "see" conflicting information; individuals may prefer to remain aligned with their social group rather than confront uncomfortable contradictions within their own beliefs or values.
Marketing	The discomfort consumers feel when their purchase decisions conflict with their beliefs about a product's value or the relevance of the purchase to their self-image.	Highlights how through persuasion techniques (seller) or justification (buyer) dissonance can shape consumer behavior and decision-making.	Illustrate that people may rationalize purchases they later feel uncertain about, blinding them to alternatives that may be better suited.
Education	The conflict students experience when their existing beliefs or knowledge are challenged by new information or experiences that contradict their prior understanding.	Serves as a motivator for students to engage with material that conflicts with their previous beliefs, ultimately encouraging critical thinking and personal growth.	Emphasizes the importance of fostering an environment where challenging existing views is welcomed.
Political Science	The tension that arises when an individual's political beliefs or party affiliations are challenged by new information or behaviors from their favored party.	Points to political polarization. When confronted with information that contradicts their political identity, individuals may resort to denial or rationalization, further entrenching their biases.	The discomfort associated with encountering contradictory information can lead to an increased commitment to previously held beliefs, preventing productive dialogue and mutual understanding.

Let's briefly explore several other fields. In **behavioral economics**, cognitive dissonance explains how individuals make decisions that seem irrational from a traditional economic standpoint (Ariely, 2008). When people make poor financial choices, they may experience dissonance between their actions and their beliefs about being financially responsible. To alleviate this discomfort, they might rationalize their spending or downplay the importance of saving. For example, a person who frequently dines out might convince themselves that they deserve the luxury despite knowing it impacts their budget negatively, effectively reducing the psychological tension associated with their choice.

In the context of **religious studies**, cognitive dissonance can arise when individuals encounter contradictions between their religious beliefs and modern scientific understandings or ethical dilemmas (Beauvais & Whelchel, 2014). This dissonance can lead individuals to reinterpret religious texts or adopt selective beliefs to maintain faith while accommodating new insights. For example, a devout believer may struggle with scientific theories on evolution and may choose to adhere strictly to a literal interpretation of sacred texts, experiencing dissonance when considering conflicting evidence from scientific communities.

Within **environmental studies**, cognitive dissonance is highly relevant when individuals learn about environmental concerns but do not change their behaviors in line with this knowledge (Steg & Vick, 2009). For example, someone may be aware of the impact of plastic use on the environment but continues to use plastic products, leading to feelings of dissonance. For example, a person may advocate for reducing plastic pollution yet frequently purchase bottled water, creating discomfort that may lead them to either justify their behavior or take steps to reconcile these conflicting beliefs.

In **philosophy**, cognitive dissonance is explored in discussions about ethical dilemmas and moral reasoning (Harman, 2011). Philosophers may examine how individuals navigate moral contradictions and the reasoning processes behind them. The discomfort caused by conflicting ethical beliefs can lead to justifications that align with one's identity or moral framework. For example, an individual may pride themselves on being environmentally conscious yet drive a gas-guzzling vehicle. The dissonance may lead them to emphasize the fuel efficiency of their vehicle or downplay the environmental impact of driving.

In the field of **media and communication studies**, cognitive dissonance is studied in relation to how media messages can challenge individuals' beliefs and attitudes (Hall, 1980). The inconsistencies between media portrayals and personal experiences can provoke dissonance, affecting how individuals consume and interpret information. For example, a In this field, cognitive dissonance is studied in relation to how media messages can challenge individuals' beliefs and attitudes. The inconsistencies between media portrayals and personal experiences can provoke dissonance, affecting how individuals consume and interpret information.

Cognitive dissonance is clearly a multifaceted concept that permeates various fields, each contributing unique insights into how individuals handle conflicting beliefs, behaviors, and information. By understanding cognitive dissonance from multiple perspectives, we can better appreciate the complexities of human thought and behavior, which can further illuminate why individuals may struggle to "see" or accept realities that challenge their established views. This comprehensive understanding can also inform strategies aimed at fostering awareness and encouraging constructive dialogue across different disciplines.

The Pivotal Role of Cognitive Dissonance

Cognitive dissonance plays a pivotal role in explaining why individuals often fail to "see" what is right in front of them. The mechanisms that drive individuals to protect their self-image and align with their social groups can engender significant blind spots in thinking and understanding. Recognizing cognitive dissonance as a fundamental aspect of human psychology can lead to greater self-awareness and open-mindedness, fostering more constructive dialogues and ultimately facilitating personal and societal growth. Acknowledging and addressing cognitive dissonance helps individuals work toward overcoming biases that hinder their ability to engage with conflicting truths and perspectives.

Here's several synthesized reasons why it is so important, particularly in relation to why we cannot see what is right in front of us.

Psychological Comfort. Cognitive dissonance arises from the discomfort of holding conflicting beliefs or values. To alleviate this discomfort, individuals often cling more tightly to their existing beliefs, leading them to ignore or downplay evidence that contradicts those beliefs. This psychological mechanism provides immediate relief but often results in a distorted view of reality, contributing to "blindness" toward factual information or alternative perspectives.

Confirmation Bias. By now, we've used this term heavily throughout, and our learning about its importance is deepening. Cognitive dissonance is heavily influenced by social identity. Individuals often derive self-worth from group affiliations (e.g., political, religious, social), and any information that threatens these group norms can lead to dissonance. To avoid this discomfort, individuals may ignore clear evidence that contradicts group beliefs, further perpetuating collective blindness to realities outside their consensus.

Social Identity and Group Dynamics. Cognitive dissonance is heavily influenced by social identity. Individuals often derive self-worth from group affiliations (e.g., political, religious, social), and any information that threatens these group norms can lead to dissonance. To avoid this discomfort, individuals may ignore clear evidence that contradicts group beliefs, further perpetuating collective blindness to realities outside their consensus.

Behavioral Implications. Cognitive dissonance impacts decision-making and behavior after the fact. For instance, once a person makes a significant choice, such as purchasing a product or supporting a political candidate, they may experience dissonance if they encounter negative information about their choice. To mitigate this discomfort, they often rationalize their decision rather than reevaluating their choice based on new evidence. This can lead to persistent blind spots regarding the quality or morality of their decisions.

Educational and Growth Challenges. Cognitive dissonance serves as both a barrier and a potential catalyst for growth. It can prevent individuals from recognizing inaccuracies in their beliefs or developing critical thinking skills, leading to resistance to new ideas. However, if people confront their dissonance constructively, it can also spark personal transformation and a reevaluation of deeply held beliefs.

Let's look at the current political climate in the U.S., which highlights the dynamics of cognitive dissonance at play, particularly around issues like the dismantling of safety nets such as Medicaid, Social Security, and various aid programs. Here's how cognitive dissonance helps explain the phenomenon of citizens ignoring or rationalizing the impacts of such political actions.

[*Attachment to Beliefs and Ideologies*] Many individuals strongly identify with specific political parties or ideologies that promote certain policies regarding social safety nets. Cognitive dissonance occurs when actions taken by their preferred politicians, such as cuts to these essential programs, conflict with their beliefs about fairness and support for vulnerable populations. Rather than facing this discomfort, people may choose to downplay or ignore the implications of these actions, prioritizing their political allegiance over critical evaluation of the policies.

[*Selective Exposure and Confirmation Bias*] Media consumption plays a significant role in this context. People often seek out news sources that align with their political beliefs, leading to selective exposure. This creates an echo chamber where dissonant information about the impacts of safety net reductions is dismissed or ridiculed. As a result, individuals may not encounter trustworthy information that challenges their views, reinforcing their blindness to the realities of those affected by aid program cuts.

[*Social Identity and Peer Influence*] The social identity aspect of cognitive dissonance is particularly palpable in political contexts. When members of a community or social group hold certain views about the government, individuals may conform to these prevailing opinions to maintain group cohesion and acceptance. This desire for belonging can result in the collective dismissal of evidence regarding the negative consequences of reducing safety nets, as doing so would create dissonance with group beliefs.

[*Rationalization of Inaction*] For many, facing the reality of weakened safety nets can be unsettling. Cognitive dissonance allows individuals to rationalize inaction; they might convince themselves that the changes are not as severe or that those affected will find support elsewhere. This rationalization serves to alleviate discomfort while maintaining their political stance, ultimately leading to apathy regarding essential social programs.

[*Emotional Resistance to Complicated Issues*] The emotional weight of social safety net issues can lead to cognitive dissonance. Many people believe in the importance of providing support to those in need, but when faced with political actions that undermine these systems, they experience discomfort. Instead of addressing the complexity of these issues, which may involve confronting uncomfortable truths about their political affiliations or responsibilities, individuals might disengage or become defensive about their beliefs.

The interplay of cognitive dissonance in the current U.S. political environment illustrates how difficult it can be for individuals to acknowledge the implications of policy changes affecting social safety nets. The desire to remain aligned with political identities and the avoidance of emotional discomfort contribute to a significant blindness regarding the harms inflicted on vulnerable populations. By uncovering these mechanisms, we can encourage more open dialogue and a reevaluation of values that prioritize welfare and social responsibility, ultimately fostering a more informed and empathetic approach to political decision-making.

Physical, Mental, Emotional and Spiritual Effects

Cognitive dissonance can have significant serious ramifications across various dimensions of an individual's life, leading to physical, mental, emotional, and even spiritual issues. Let's briefly look how cognitive dissonance can manifest and affect these different aspects of human experience.

Physical Effects. While cognitive dissonance primarily impacts psychological processes, it can also lead to physical consequences. In terms of *stress responses*, the discomfort caused by holding conflicting beliefs often triggers the body's fight-or-flight response, resulting in chronic stress. Physical symptoms may include headaches, muscle tension, fatigue, and elevated heart rates. Related to *health compromises*, individuals dealing with unresolved dissonance may neglect healthy lifestyle choices or self-care practices, leading to issues such as poor diet, lack of exercise, or insufficient sleep, which can contribute to long-term health problems.

Mental Effects. Cognitive dissonance profoundly affects mental well-being. There is *reduced cognitive function*. The strain of holding contradictory beliefs can lead to decreased focus and mental clarity. Individuals may find it challenging to process new information or engage in critical thinking. It causes *confusion and indecision*. People experiencing cognitive dissonance may struggle with decision-making, feeling caught between conflicting choices. This can lead to analysis paralysis, where individuals are unable to make decisions for fear of further dissonance. Constantly experiencing cognitive dissonance can overwhelm individuals mentally, leading to *cognitive overload*. Individuals may struggle with processing information, leading to burnout or disengagement from both personal and professional endeavors. Prolonged exposure to dissonance can contribute to mental health issues such as depression or chronic stress, especially if individuals feel they cannot resolve the conflict.

Emotional Effects. Cognitive dissonance can significantly impact emotional health. It can cause *anxiety and depression*. The internal turmoil created by dissonance can heighten anxiety and lead to depressive symptoms. Individuals may feel a sense of hopelessness or helplessness, particularly if they believe they can't reconcile their conflicting beliefs. It can cause *feelings of guilt and shame*. When personal values conflict with behaviors, individuals may experience guilt or shame, particularly if they engage in actions they believe to be immoral or harmful. This can further exacerbate emotional distress. When cognitive dissonance is induced manipulatively, as in some political or interpersonal scenarios, individuals may *experience resentment* towards the person or entity causing the discomfort. This can lead to distrust and a breakdown in relationships. Such negative feelings can diminish cooperation and dialogue, creating hostility and polarization rather than fostering understanding or consensus.

Spiritual Effects. The impact of cognitive dissonance can extend into spiritual or existential domains. It can facilitate a *crisis of faith*. For individuals with strong religious or spiritual beliefs, encountering dissonance between their faith and new evidence or experiences can lead to spiritual crises. This may cause them to question the validity of their beliefs or grapple with profound uncertainty. It may be accompanied by a *loss of purpose*. Dissonance can challenge one's sense of identity and purpose. When beliefs about self-worth, morality, or meaning are in conflict, individuals may struggle to find direction or fulfillment in their lives. Cognitive dissonance often threatens *personal or group identity*, particularly when individuals strongly associate their beliefs with their self-concept. Being confronted with dissonance can lead to an identity crisis or existential anxiety. This threat can provoke feelings of inadequacy and defensiveness, hampering personal growth and leading to conflict with others who hold differing views

Defensive reactions cause individuals to double down on their existing beliefs or become even more entrenched in their positions to avoid confronting the dissonance. This can create an echo chamber effect, where rather than reconsidering beliefs, individuals may reinforce their viewpoints, thus intensifying polarization and reducing constructive communication.

As can be seen, cognitive dissonance is more than a benign psychological phenomenon; it can have serious and far-reaching effects on physical, mental, emotional, and spiritual health. The discomfort of conflicting beliefs can lead to stress, anxiety, and a re-evaluation of one's values and identity. Understanding and addressing cognitive dissonance is essential not only for personal growth and development but also for overall well-being. By fostering self-awareness and open-mindedness, individuals can work toward resolving dissonance, leading to greater alignment between their beliefs, values, and actions.

It Can Be Purposefully Caused by Others

Cognitive dissonance can be purposefully induced by others, often used as a *means of persuasion or manipulation*. This deliberate creation of dissonance can occur in various contexts, including advertising, politics, education, and interpersonal relationships. Let's explore ways in which others might intentionally cause cognitive dissonance in terms of advertising and marketing, political messaging, persuasive communication, psychological manipulation, and educational strategies, and its potential result.

In **advertising and marketing**, advertisers often create dissonance to encourage consumers to buy their products. This is done by showcasing the advantages of a product while hinting at the downsides of not using it. For example, a marketing campaign may highlight how using a particular eco-friendly product contributes to environmental protection, thus generating discomfort for those who continue to use less sustainable alternatives.

For **political messaging**, politicians and political organizations may exploit cognitive dissonance to sway public opinion. They might present information that challenges established beliefs or values among their constituents, encouraging individuals to reconsider their viewpoints or align with specific narratives. For example, if a politician advocates for reforms that threaten a core belief of their supporters, they may frame the discussion in a way that creates dissonance and encourages the audience to reconcile their beliefs with a new political reality.

To achieve **persuasive communication** in debate or negotiation settings, individuals may intentionally highlight inconsistencies in the other party's beliefs or behaviors. This tactic can create discomfort, pushing the other party to reevaluate their stance or concede to pressure. For example, a manager may point out that an employee's actions contradict the company's stated values to encourage the employee to align their behaviors with organizational expectations.

As part of educational strategies, educators may sometimes use cognitive dissonance as a **teaching method**, presenting new information that conflicts with students' preconceived notions. This approach can stimulate critical thinking and encourage learners to adjust their understanding based on new evidence. While the intention is often constructive, it can still create discomfort, which can be an effective way to promote learning and personal growth.

It can also be used for **psychological manipulation**. In some interpersonal relationships, cognitive dissonance can be purposefully induced through gaslighting or other manipulative tactics. For instance, an individual might consistently contradict another's beliefs or perceptions, leading the latter to question their reality and ultimately align their beliefs with the manipulator's. This creates cognitive dissonance, as the victim struggles to reconcile the conflicting evidence presented by the manipulator with their experiences.

Regardless of the setting, intentionally inducing cognitive dissonance for manipulative purposes, whether psychological manipulation or gaslighting, can lead to serious violations of trust and autonomy. This can have long-term detrimental effects on the victim's self-esteem and self-concept. Victims may develop conditions like Stockholm syndrome or experience difficulty in forming healthy relationships in the future, as their ability to trust may be compromised.

While inducing cognitive dissonance is often employed as a strategic tool for persuasion or influence with benefits accrued to those using it, this comes with ethical implications. Intentionally creating emotional and psychological discomfort in others can result in harm and mistrust if not handled with care. Understanding these dynamics can help individuals navigate situations where they may be subjected to cognitive dissonance, ultimately allowing them to make more informed decisions regarding their beliefs and actions.

Looking Through the Lens of Cognitive Dissonance

The current administration in the U.S. has employed various communication strategies that can be analyzed through the lens of cognitive dissonance. While it's essential to clarify that specifics might vary in execution and intention, cognitive dissonance can exist in the following ways within political discourse and policy-making.

[*Challenging Established Norms*] The administration may present policies or narratives that directly challenge widely held beliefs or expectations among constituents. By pushing for changes that contrast with traditional viewpoints (e.g., climate change initiatives or healthcare reforms), the administration can induce cognitive dissonance among those who might typically oppose such measures. This dissonance can prompt constituents to either reconsider their positions or rationalize their allegiance to the administration by aligning with its narrative.

[*Framing and Messaging*] Political messaging often aims to create a sense of urgency or necessity around certain issues, highlighting the supposed consequences of inaction. For example, framing discussions about the economy, healthcare, or social justice in a way that emphasizes impending crises can induce dissonance in individuals who may otherwise be comfortable with the status quo. The administration may seek to create empathy for marginalized groups or pressures in healthcare systems, leading some to reconsider their beliefs about government intervention.

[*Promoting Contradictory Information*] At times, administrations deliberately introduce information that contradicts pre-existing beliefs among constituents to stimulate discussion or mobilize action. For example, addressing misinformation around vaccination or public health measures can create dissonance in those who distrust government recommendations, pushing them to reevaluate their beliefs in light of new evidence. Or, conversely, it can be used to promote confusion, reducing support for ideas to which the administration is opposed.

[*Strengthening Political Loyalty*] By inducing cognitive dissonance among constituents—particularly those aligned with certain political ideologies—the administration may reinforce loyalty. When supporters are confronted with policies that create discomfort but are framed in terms of greater good or moral imperative, they may rationalize their support, ultimately solidifying their allegiance to the administration.

While the administration may not explicitly state that they are employing cognitive dissonance as a strategy, as can be seen many political communication techniques can inherently induce this psychological phenomenon. The goal is often to reshape public perception, mobilize support, and challenge entrenched beliefs. Understanding these dynamics allows constituents to recognize how political discourse can influence thought processes and perceptions, fostering critical engagement with policies and narratives presented by the government.

The Dissonance Within (A Story)

Setting: The small town of Harmonyville, nestled in the heart of America, where the political climate has turned tense. Residents are divided over government reforms aimed at cutting safety nets like Medicaid, Social Security, and public aid programs. As the town prepares for an upcoming election, the consequences of these reforms loom large, yet many are blissfully unaware.

Characters:

1. **Maggie**: A single mother working two jobs, both reliant on Medicaid for her health and her children's wellbeing.
2. **Tom**: An ardent supporter of the current administration, an optimistic small business owner who believes in "personal responsibility."
3. **Evelyn**: A retired schoolteacher who has relied on Social Security; she holds strong convictions about the importance of caring for the vulnerable.
4. **Marcus**: A college student studying sociology, who aims to understand the impacts of the political decisions being made around him.

As the sun rose over Harmonyville, the town buzzed with anticipation for an upcoming town hall meeting to discuss the proposed cuts to social safety net programs. The atmosphere was charged with conflicting opinions, and no one felt the tension more than Maggie, who had just finished a grueling night shift.

[Scene 1: A Mother's Concern]

Maggie sat with her children at the breakfast table, her mind racing. She had spent the last few months juggling her jobs, often pulling double shifts to keep up with expenses. The thought of losing her Medicaid coverage unsettled her deeply.

Maggie: "I don't know what I'll do if they cut Medicaid. I can barely afford rent as it is!"

Her oldest daughter, Lucy, looked up from her cereal.

Lucy: "But Mom, people say the government just wants to help us be more efficient, to cut costs. Maybe it'll be okay?"

Maggie frowned, feeling a sense of cognitive dissonance. She wanted to believe that her government was looking out for her, yet her reality told a different story. She tucked her worries away and forced a smile for her children, clinging desperately to her belief in the system.

[Scene 2: Tom's Optimism]

On the other side of town, Tom was readying his small café for the day's customers. He had been listening to the news, which praised the government's efforts to promote personal responsibility and reduce what they called "wasteful spending."

As he served coffee to a group of loyal regulars, he couldn't help but express his views.

Tom: "You know, cutting down on these safety nets will encourage people to get out there and work harder! It's about time we stop coddling them."

His words resonated with many customers, but Evelyn, a long-time patron, felt a pang of discomfort. She had just turned 70, counting on her Social Security to make ends meet.

Evelyn: "Tom, you've never had to depend on those programs. Some of us do."

[Scene 3: Tension Builds]

At the town hall meeting that evening, the room filled quickly. Residents filled the seats, some in anticipation, while others came armed with their perspectives. Maggie walked in nervously, clutching the back of her chair. Tom, seated near the front, greeted her with a wave.

The meeting started with the mayor presenting statistics that highlighted the government's successes. Maggie felt her heart racing as she listened to rationale that sounded good on paper but didn't match her reality.

When questions opened from the audience, Evelyn raised her hand.

Evelyn: "What about the impact of cutting Social Security? For many of us, it's the difference between eating and going hungry."

Tom's expression shifted to discomfort. He hadn't considered the tangible effects of the policies.

Tom "But isn't it better for people to seek jobs rather than rely on the government?"

The room tensed. Maggie couldn't hold back.

Maggie: "I work two jobs! I rely on Medicaid because I simply can't afford insurance. This isn't just about politics—it's about real lives!"

[Scene 4: Confronting Dissonance]

As the discussion grew heated, Marcus, who had been quietly taking notes at the back, felt compelled to speak.

Marcus: "We're caught in this cycle of beliefs. It's okay to question our views! Just because it challenges what we believe doesn't make it wrong."

His voice cut through the tension, prompting murmurs of agreement and dissent.

[Scene 5: Reflection and Resolution]

The crowd began to engage in a stronger discussion. Maggie, Tom, and Evelyn found themselves discussing the issues more openly, voices raised but with newfound respect.

Tom, feeling his own cognitive dissonance, began to reconsider some of his views. The fear of losing community support was no longer stronger than the empathy he felt listening to Maggie and Evelyn.

Tom: "You're right. Maybe we need to think beyond our beliefs and consider how these changes impact those who truly rely on them."

As the meeting wrapped up, there was an air of uncertainty but also hope. The townspeople began to realize that open dialogue could bridge divides. While everyone wouldn't agree, acknowledging each other's realities became a powerful step toward understanding.

In the coming weeks, Harmonyville would not change overnight, but the town would begin to recognize the impact of cognitive dissonance on their perspectives. They learned to confront their discomfort, engage with opposing views, and strive for a community where every voice mattered. With each conversation, they peeled back layers of blindness, inching closer to seeing the truth behind their beliefs, and recognizing that which was right in front of them but they previously could not see.

Chapter 12
Self-Deception

In the mirror of self, we may hide
Where truths and illusions collide.
Whysly we pry,
Unravel the lie,
To see past the deceptions inside.

Self-Deception is crucial to understanding why we often fail to "see what is right in front of us," especially if we interpret "see" broadly to mean recognize, understand, acknowledge, or know. Before diving into the concept of self-deception and connecting it to the learned understanding leading up to this chapter, let's clarify our understanding of the concept of *self*.

The Concept of Self

Originating from the Proto-Indo-European root *sel-bho-* meaning "self, by oneself," in Old High German it became *selb* and the Gothic *silba*, morphing into Old English as *seolf* or *self*, all meaning roughly the same as the modern English translation. Consulting the *Oxford English Dictionary* and *Oxford Languages*, the concepts of self can be described as:

(1) A person's essential being that distinguishes them from others, especially considered as the object of introspection or reflexive action. *Example:* She found herself gain after years of searching.

(2) One's own interests, welfare, or advantage. *Example:* He acted out of self.

(3) The qualities that make a person or thing unique; identity or individuality. *Example:* He is finally his true self.

In summary, self is the conscious, reflective identity or being of a person, regarded as distinct from others and as the subject of their own experiences, actions, and emotions. That's a pretty good description.

Each and every human being is unique, whether considered from the frame of reference of DNA, experience, culture, family, thought patterns, beliefs, values, or emotions and feelings. At the core of all this difference is the *self*, with a subjective mind, exploring the world from the inside out, a protagonist ready for action. Yet in the magical environment of the womb, self does not exist. The thinker of the future that we become (the self) is a blank slate (Bennet, Bennet, & Turner, 2018).

Self is an emergent quality of the human which moves beyond biological drivers and cultural habits, the totality of the conscious and unconscious mind, the brain and the body. Consistent with the sharing forwarded in this book, Csikszentmihalyi (1993) describes this when he says, "Inside each person there is a wonderful capacity to reflect on the information that the various sense organs register, and to direct and control these experiences" (p. 22). This is rather like a figment of our imagination, "something we create to account for the multiplicity of impressions, emotions, thoughts, and feelings that the brain records in consciousness" (p. 216).

Self—including self-referential memory, self-description, self-awareness and the personality—coevolve with, and is the creator of, its environment, which is a process not strictly within an organism but *between* the organism and its environment (Bennet, Bennet, & Turner, 2018). We often perceive ourselves as the roles we play, the identities we take on, but those are just the *way we choose* to manifest self. Self is so much more.

Contemporary neuroscience does not identify a separate neurological function or structure where self—or consciousness, which enables the recognition of self—exists. Self is the Quantum leap that occurred after the emergence of self-reflective consciousness, a distinct self that could take charge of the domain of consciousness, and determine which feelings or ideas take precedence. As Csikszentmihalyi (1993, p. 23) concludes: "Having had this experience of something inside us directing consciousness we gave it a name—the self—and took its reality for granted. And the self became an increasingly important part of human beings."

What is Self-Deception?

Self-deception is the process by which we convince ourselves of a truth or falsity, often unconsciously, despite clear evidence to the contrary. It's not *simply* lying to ourselves—it involves a complex mental dance where the mind both "knows and does not know" something at the same time.

Three key mechanisms engaged in self-deception, some of which have been discussed in detail in previous chapters, are motivated blindness/motivated reasoning; selective attention/confirmation bias; and cognitive dissonance and divided mind. We'll just do a short review here, focusing on the areas that new to our discussion.

Motivated Blindness/Motivated Reasoning. Motivated reasoning theory posits that people's desires and preferences influence their beliefs and interpretations of evidence (Sharot, 2011; Kunda, 1990). Motivated reasoning explains how people unconsciously filter data so that they only "see" (recognize or accept) what aligns with their desires or beliefs, and ignore or reinterpret the rest. This cognitive filtering is why we can fail to recognize truths "right in front of us" that would be psychologically threatening or uncomfortable. For example, a person invested in their job may overlook clear signs their company is failing, because accepting the truth would mean facing insecurity and loss.

Selective Attention/Confirmation Bias. Attentional bias theories describes how people unconsciously focus on information that confirms what they already believe. Selective attention means we literally don't "see" contradictory evidence, or we see it but quickly disregard it, so it never enters our awareness in a meaningful way. Our perception is shaped by our expectations and biases, keeping inconvenient facts out of view even if they are "right in front of us." For example, a sports fan might notice every referee call against their own team but ignore (not "see") equally questionable calls favoring their team.

Cognitive Dissonance and Divided Mind. As previously discussed, cognitive dissonance theory holds that we experience discomfort when our beliefs and evidence conflict; to resolve this, we may change our beliefs or simply block out the conflicting information.

Dissonance theory and existential philosophy see self-deception as a kind of splitting: part of the mind "knows" something, while another part keeps it at bay (often unconsciously). Philosophers such as Satre and Nietzsche and psychologists like Freud have described this as a mind partitioning itself, with one "part" hiding knowledge from another. This internal division explains how we can remain functionally blind or in denial of clear realities, unable to "see" what's right before us because we are unconsciously invested in not knowing.

Here's a quick summary with the supporting theory.

Mechanism	Theory and Reference	How it blocks "seeing"
Selective Attention	Confirmation Bias (Nickerson, 1998; Festinger, 1956)	Noticing only what fits beliefs
Motivated Blindness	Motivated reasoning (Kunda, 1990; Sharot, 2011)	Filters out pain/undesirable truths
Divided Mind / Dissonance	Cognitive Dissonance	Keeps knowledge out of awareness.

These frameworks—motivated reasoning, selective attention/ confirmation bias, and cognitive dissonance/splitting—are the leading theoretical tools explaining self-deception in philosophy and psychology. They show how, through both unconscious and motivated processes, we can fail to recognize, acknowledge, and act on truths that are literally and metaphorically "right in front of us."

How Did Self-Deception Evolve?

The persistence and universality of self-deception invite a deeper question: If self-deception often leads to poor decisions or blindness to reality, why did such a tendency evolve in human beings? Far from being a quirk or flaw, leading evolutionary theorists propose that self-deception may serve important adaptive functions, both for individuals and societies.

Perhaps the most influential evolutionary account comes from biologist Robert Trivers (2000), who argued that self-deception evolved primarily as **a means of better deceiving others**. In his seminal paper, "The Elements of a Scientific Theory of Self-Deception," and later in *The Folly of Fools*, he writes: "Self-deception evolves in the service of deception—the better to fool others. The primary logic is: it is easier to fool others if you believe the lie yourself" (Trivers, 2011, p. 8).

This logic suggests that by sincerely believing a falsehood, an individual can avoid the telltale signs of conscious lying—hesitation, microexpressions, inconsistencies—thus increasing their social and reproductive advantage. Because humans are highly social and rely on trust, the ability to evade detection while attempting to manipulate others (even unconsciously) could provide significant evolutionary benefits in competition for resources, mates, and status (von Hippel & Trivers, 2011).

Another set of evolutionary hypotheses views self-deception as **a buffer against psychological pain or cognitive overload**. Believing in one's competence or safety (even in the face of contrary evidence) may help individuals (2) sustain confidence and motivation in adversity (Johnson & Fowler, 2011); and (2) reduce paralyzing anxiety or fear that could otherwise prevent necessary action (Nesse, 1990). For instance, overoptimism and denial may help a person attempt risky but potentially life-saving or reproductively beneficial activities—such as hunting, courting, or entering conflict. In many ancestral situations, overestimating risk or danger could lead to missed opportunities, so those capable of self-deception may have enjoyed fitness advantages.

Self-deception can also operate at the group level by **promoting cohesion and shared narratives**. Cultures transmit myths, origin stories, and simple moral binaries that, while often factually questionable, help coordinate pro-social behavior and collective action (McKay & Dennett, 2009). Individuals who self-deceive about their group's superiority or righteousness may behave more loyally and bravely, conferring collective benefits despite personal misjudgments (Kurzban & Aktipis, 2007).

Evolution may also favor mechanisms that help humans resolve conflicting beliefs, values, or information, thus reducing cognitive dissonance (Festinger, 1957). When caught between contradictory needs (e.g., self-preservation vs. group loyalty), self-deception enables swift decision-making and action, preventing "analysis paralysis."

Here's a summary table:

Adaptive Function	Description	Key References
Deceiving others ("liar and believer as one"	Aids social manipulation, increases trust from others	Trivers (2000, 2011); von Hippel & Trivers (2011)
Emotional buffering / confidence	Sustains motivation, reduced stress and discouragement	Johnson & Fowler (2011); Nesse (1990)
Group cohesion	Enables cooperation and collective identity, simplifies social dynamics	McKay & Dennett (2009); Kurzban & Aktipis (2007)
Cognitive conflict resolution	Reduces mental distress from conflicting beliefs or values	Festinger (1957); Trivers (2000)

Thus, self-deception is seen not as a cognitive failing, but as representing an evolved repertoire of strategies for navigating a complex social world. It can help us convince others—and ourselves—of a chosen story, motivate action in uncertainty, unite groups, and reduce psychological distress. However, as contemporary contexts differ from ancestral ones, these once-adaptive tendencies may now have ambiguous or maladaptive consequences

The Way Individuals Make Choices

In today's environment, self-deception plays a significant role in the way individuals make choices, influencing both decision-making processes and outcomes. Why is this true? Why would we deceive our self? The answer is multifold, and directly related to the core theme of this book. Here are several key aspects of how self-deception affects choice.

Cognitive Dissonance. Recall that cognitive dissonance refers to the mental discomfort experienced when holding two conflicting beliefs or attitudes. As is clear from the previous chapter, self-deception can serve as a coping mechanism to reduce or eliminate this discomfort by justifying inconsistent behaviors. For example, if someone believes they should exercise but doesn't, they might deceive themselves into thinking that they are too busy or that exercise isn't that important, rather than confronting the reality of their choices.

Motivation and Confidence. By deceiving themselves about their abilities or the likelihood of success, individuals can bolster their confidence and motivation. Self-deception can create a sense of inflated confidence, which may motivate individuals to pursue goals they might otherwise shy away from. This temporary boost can lead people to take on challenges with a roe positive outlook, potentially leading to better outcomes, or it can encourage people to take on challenges that they are unprepared to face (Leman & Cinnirella, 2007). For instance, an entrepreneur may overlook potential risks of a new venture to maintain optimism.

Risk-Taking. Self-deception may lead individuals to underestimate risks associated with their choices, promoting bolder decision. This can be beneficial in situations requiring decisiveness but could also lead to reckless behavior. This phenomenon is often observed in environments such as gambling or investing, where ignoring potential downsides may seem beneficial (Dunning, Heath, & Suls, 2004).

Avoiding Responsibility. Self-deception can provide an escape from accountability, allowing individuals to blame external circumstances or situational factors rather than take responsibility for their choices. By rationalizing a poor decision or blaming external factors, individuals might avoid facing the consequences of their actions, hindering personal growth (Baumeister & Exline, 2000).

Emotional Regulation. It can serve as a coping mechanism, allowing people to manage anxiety or fear associated with difficult decisions. Self-deception might allow someone to downplay the stakes involved in a choice, allowing them to avoid feelings of anxiety or fear that come with difficult choices, and making it easier to move forward. This mechanism also reduced stress (Dunning & Story, 1991).

Social Relationships. In interpersonal contexts, self-deception may shape choices around relationships. For example, ignoring red flags or toxicity, individuals may convince themselves that a partner is more compatible than they actually are, impacting relationship decisions. This can result in choices that perpetuate unhealthy dynamics.

Examples: (Relationships) Not seeing someone's lack of love or investment, despite obvious signs, because we don't want to acknowledge pain or loss. (Health) Ignoring symptoms, downplaying risk factors, or not seeking diagnosis, because the reality is too scary. (Prejudice) Failing to see our own biases because admitting them would challenge our self-image as good or fair-minded.

Overall, while self-deception can facilitate decision-making and mask uncomfortable truths, it may also lead to poor choices and negative consequences in the long run. Balancing self-awareness with the instinct to deceive oneself is crucial for healthier decision-making.

Self-deception isn't just an individual quirk—it's a universal, often adaptive human trait, which is *how and why our minds can keep us from recognizing, accepting, or acting on, obvious facts right in front of us*. It's a protective—but double-edged—mechanism that shapes our own blind spots. It helps us maintain hope, reduce anxiety, or persist in adversity. But it can also trap us in denial, make us blind to important truths, and prevent growth or problem-solving.

When Protection Becomes a Trap

While self-deception often serves adaptive functions—protecting us from psychological pain, preserving motivation, or easing social bonds—its protective veil can all too easily become a trap. **Maladaptive self-deception** occurs when the refusal to see or acknowledge reality leads to outcomes that are harmful both to individuals and groups.

Self-deception shifts from adaptive to maladaptive when (1) it persistently blocks accurate self-assessment or feedback; (2) it prevents learning from error or leads to direct harm; or (3) when the cost of "not knowing" outweighs the protective benefit. Recognizing these limits is essential. The same shield that preserves us in small doses may imprison us if worn too long or too tightly.

Personal consequences to maladaptive self-deception are multiple. In the area of health, ignoring troubling symptoms or rationalizing unhealthy behaviors can delay necessary medical care, worsening illnesses that might have been treatable (Sackeim & Gur, 1978). It can also affect relationships. Downplaying red flags or one's own destructive habits may sustain toxic relationships but costs emotional health and growth (Baumeister & Exline, 2000). Relative to addiction, many people with substance use problems minimize their dependence ("I can quit anytime") or rationalize continued use, making recovery more difficult (Dunning, Heath, & Suls, 2004). And it can stunt self-improvement. By refusing to see their own weakness or error, individuals forfeit opportunities to change, learn, or repair.

There are also collective dangers. For example, potential organizational catastrophe. By refusing to see their own weakness or error, individuals forfeit opportunities to change, learn, or repair. It can also promote social injustice. Societies may perpetuate harmful systems by denying inequity or rationalizing prejudice (Jost & Hunyady, 2003; Cohen, 2001). And it plays into global threats. Climate denial, racism, or disease outbreaks illustrate how collective self-deception can threaten not just groups, but humanity at large (Norgaard, 2011).

Self-Deception in Group and Collective Contexts

While self-deception is commonly discussed as an individual psychological phenomenon, it also manifests powerfully at the group or collective level. Remarkably, entire organizations, communities, and even nations—or parts of nations—can become "blind" to obvious truths, engaging in forms of collective denial or groupthink that resemble individual self-deception but scale up through social dynamics and cultural narratives. A modern day example is the MAGA movement in the United States.

One core manifestation of collective self-deception is **groupthink**, a process identified by Irving Janis (1972) in which cohesive groups strive for unanimity at the expense of critical evaluation. Members suppress dissent, ignore warning signs, and rationalize poor decisions in order to maintain harmony and a shared illusion of invulnerability or correctness. This can result in catastrophic outcomes—Janis famously analyzed it in contexts like the Bay of Pigs invasion—where groups collectively fail to perceive clear risks or evidence "right in front of them." As Janis (1972, p. ix) summarized: "The more amiability and esprit de corps there is among the members of a policy-making in-group, the greater the danger that independent critical thinking will be replaced by groupthink, which is likely to result in irrational and dehumanizing actions…."

At an even broader scale, societies create and maintain myths, shared beliefs, and "unspoken taboos" that foster collective self-deception. Cohen (2001) describes how states and populations manage "states of denial," selectively ignoring or minimizing knowledge of atrocities, systemic injustice, or ecological threats. Entire cultures may propagate sanitized versions of their own histories (e.g., ignoring historical atrocities or inconvenient episodes) or downplay ongoing social problems because fully acknowledging these realities would threaten group identity, power structures, or existential security.

Collective self-deception may take several interrelated forms. First, *shared confirmation bias*. Groups reinforce and amplify shared beliefs, selectively attending to information that confirms the in-group's worldview while dismissing dissent or contradictory facts (Sunstein, 2001). For example, for years, many governments and populations ignored or minimized the scientific consensus on climate change, engaging in collective denial despite increasingly obvious evidence (Norgaard, 2011).

Second, *social reinforcement*. Individuals may privately doubt the collective narrative but go along due to social pressure, conformity, or fear of ostracism, thereby perpetuating the group's self-deception (Festinger et al., 1956; Asch, 1955). For example, societies often downplay, rationalize, or actively forget troublesome aspects of their own histories to maintain a positive collective self-image (Cohen, 2001; Zerubavel, 1995).

Third, *authority and propaganda*. Leaders or powerful institutions can shape "official stories" that become near-impossible to question within the group, institutionalizing collective self-deception (Jost & Hunyady, 2003). For example, corporations or bureaucracies may collectively ignore signs of misconduct or risk, as illustrated by cases like Enron or the Challenger disaster (Vaughan, 1996).

Conversely, from an evolutionary or social-functional standpoint, some degree of collective self-deception may have conferred adaptive benefits such as cohesion. Shared narratives, even if false or idealized, can foster group cohesion, morale, and a sense of purpose (Kurzban & Aktipis, 2007). A second example would be related to action in crisis. Shared narratives, even if false or idealized, can foster group cohesion, morale, and a sense of purpose (Kurzban & Aktipis, 2007).

However, collective self-deception can be profoundly dangerous, leading to missed threats, propagation of injustice, or perpetuation of catastrophic errors. Here's a summary table:

Phenomenon	Collective Effect	Key References
Groupthink	Suppresses dissent, amplifies shared false beliefs	Janis (1972); Sunstein (2001)
States of denial	Societal-level avoidance of uncomfortable truths	Cohen (2001); Festinger et al., (1956)
Historical amnesia	Sanitizing or erasing troubling aspects of group history	Zerubavel (1995); Cohen (2001)
System justification	Rationalizing and perpetuating flawed social systems	Jost & Hunyady (2003); Norgaard (2011)

Just as the individual mind protects itself through self-deception, so too can groups and societies. The mechanisms are strikingly similar: resistance to threatening truths, selective attention, and the need for cohesion and identity. Yet the consequences of collective self-deception—war, environmental destruction, persistent injustice—remind us that seeing what's "right in front of us" is as much a societal challenge as it is a personal one

The Blue Tower's Blind Spot (A Story)

For thirty years, the Blue Tower reigned over the city's skyline—a gleaming corporate headquarters, home to hundreds of ambitious employees. Inside, the executive team prided themselves on their open-door policy and culture of innovation. Every quarter, they gathered in the glass-walled conference room to review numbers, brainstorm new projects, and savor the Tower's success.

Yet in recent years, cracks had appeared in the company's foundation. The market was shifting, younger startups were gaining ground, and customer complaints were rising. Early warning signs trickled in—an analyst's report flagged declining sales in the Northeast sector; a junior technician sent an email about server vulnerabilities; a seasoned manager quietly requested a meeting to express her concerns about morale. Each concern, on its own, seemed minor. Each was, subtly, dismissed. The executive team told themselves—and each other—that downturns were temporary. "We've weathered storms before," said the CEO, and heads nodded in agreement. They repeated their founding mantra: "The Tower Stands Firm."

In the lunchroom, employees joked about "Tower Logic"—the way feedback disappeared in suggestion boxes, or how bad news seemed to bounce right off the marble floors. New hires quickly learned to echo positive refrains. No one wanted to be the sour voice in a chorus of optimism.

Months passed, then years. Competitors outpaced them; clients drifted away. Still, the quarterly meetings maintained their routine, complete with slide decks spinning setbacks as "transitional opportunities." The head of marketing muttered privately to a friend, "Are we the only ones who see what's happening?" But even he, eventually, stopped raising his worries in meetings—after all, every conversation circled back to the same hopeful scripts.

It was only when the layoffs began—first dozens, then hundreds—that the truth became unavoidable. Angry shareholders demanded change. The CEO, shaken, looked over the last ten years of reports and memos, and was staggered by what he saw: so many warnings had gone unread, so many uncomfortable facts had been politely reworded, then quietly set aside. The Tower, it turned out, was full of eyes that had refused to see.

Reflecting later, a former manager would recall: "It was all right in front of us—the numbers, the trends, the tension in the meetings—and yet, together, we just… didn't see. We couldn't, because seeing it hurt too much. So we told each other the story we needed to hear, and called it the truth."

Moral: In the Blue Tower, group self-deception grew not from any single lie, but from subtle habits of collective denial. Shielded by shared optimism, each person ignored the obvious—until reality broke through. As in many real organizations, the greatest danger was not that no one knew, but that everyone, together, refused to see what was right in front of them.

What Fields of Research Explore Self-Deception?

Self-deception primarily falls under the field of psychology, particularly within the domains of cognitive psychology and social psychology. Cognitive psychology examines the mental processes underlying self-deception, such as memory, perception, and reasoning. Researchers study how cognitive mechanisms can lead individuals to deceive themselves, including how they process information to maintain a positive self-image.

In social psychology, self-deception is explored in the context of interpersonal relationships and social behaviors. This includes how individuals may deceive themselves to maintain social bonds, manage impressions, or justify actions that may conflict with their moral or ethical standards.

Self-deception is also relevant in other areas. For example, philosophers investigate the nature of self-deception, its ethical implications, and its relationship to concepts like truth and integrity. In behavioral economics,, self-deception is examined regarding decision-making and how individuals may rationalize poor financial choices. In clinical settings (psychotherapy), understanding self-deception is important for addressing personal challenges and promoting self-awareness.

Overall, self-deception is a complex phenomenon that intersects various disciplines, highlighting its significance in understanding human behavior and thought processes.

Can We Overcome Self-Deception?

Recognizing that self-deception is deeply woven into our psychology and cultures raises a crucial challenge: *Can we actually overcome these blind spots and see ourselves (and reality) more clearly?* The answer, according to research, is "yes—but not easily or entirely."

Self-deception operates largely outside of conscious awareness (Trivers, 2011), and many of its cognitive roots are automatic. Yet, a variety of psychological and philosophical traditions—from psychotherapy and mindfulness to ancient wisdom—suggest that self-awareness can be increased, and self-deception reduced, through systematic effort.

Research and practical wisdom point to several strategies that can help loosen the grip of self-deception, most of which you will be familiar with.

These are mindfulness and present-moment awareness, seeking honest feedback, reflection and journaling, welcoming discomfort and dissonance,, challenging cognitive biases, and therapeutic and group approaches. These are briefly addressed below.

Mindfulness and Present-Moment Awareness. Practicing nonjudgmental awareness of thoughts, feelings, and bodily sensations makes it easier to notice self-deceptive patterns as they arise, rather than running on autopilot (Brown, Ryan, & Creswell, 2007).

Seeking Honest Feedback. Trusted friends, mentors, or counselors can help us see what we overlook or rationalize, offering external perspectives that counter our biases (Vazire & Carlson, 2011).

Reflection and Journaling. Regularly examining our motives, assumptions, and past decisions—especially in writing—can reveal patterns of avoidance or rationalization we might otherwise miss (Pennebaker, 1997).

Welcoming Discomfort and Dissonance. Instead of avoiding or suppressing feelings of discomfort when confronted with contrary evidence, facing this discomfort and questioning our defensive reactions can signal areas ripe for growth (Festinger, 1957).

Challenging Cognitive Biases. Actively questioning our own beliefs, entertaining opposing viewpoints, and scrutinizing "too convenient" explanations can reduce confirmation bias and motivated reasoning (Lilienfeld, Ammirati, & Landfield, 2009).

Therapeutic and Group Approaches. Cognitive Behavioral Therapy (CBT), group dialogues, and self-inquiry techniques are evidence-based methods for surfacing, examining, and challenging self-deceptive habits (Beck, 2011).

It is well worth the effort to work at overcoming self-deception. Cultivating self-awareness is foundational to healthier decisions, personal growth, learning, relationships, and ethical living (Wilson & Dunn, 2004). By honestly confronting uncomfortable truths, people can break cycles of denial, learn from mistakes, and chart more authentic courses for their lives.

Consistent with these strategies, let's develop a step-by-step approach to cultivating self-awareness and reducing self-deception along with a scenario example.

Overcoming Self-Deception Step by Step

Scenario Example Context: Sam has always considered himself a loyal and easygoing friend. Lately, however, he's noticed that he and his friend Jose rarely talk unless Sam initiates. When they do meet, conversations feel awkward, and Jose often seems distracted or uninterested. Sam brushes off his concerns, telling himself Jose is just "busy," but a part of him feels uneasy.

Steps:

Step 1: Commit to Self-Inquiry. Set an intention to be honest with yourself, even when the truth is uncomfortable. Recognize that "seeing" is an ongoing practice, not a single event.

Scenario Example: Sam resolves to genuinely examine his feelings rather than brushing them aside. He acknowledges, "I feel anxious something's off with Jose. Maybe I need to look closer at what's happening, not just what I want to be true."

Step 2: Develop Mindful Awareness. Allocate time each day, even 5–10 minutes, to notice your thoughts and feelings without judgment. Techniques include meditation, body scans, or simply pausing to reflect on your emotional state.

Scenario Example: The next afternoon, Sam goes for a walk and practices noticing his thoughts and feelings, letting them surface without judgment. He realizes, "I feel hurt—and kind of afraid that maybe Jose is annoyed with me. That's uncomfortable."

Step 3: Practice Reflective Journaling. Keep a regular journal in which you not only record events, but also reflect on your reactions, decisions, and their motivations. Ask yourself: "What story am I telling myself? What might I not want to face?"

Scenario Example: That night, Sam writes, "Why do I always have to text Jose first? He hardly asks about my life lately. I keep telling myself he's just stressed, but am I ignoring signs that our friendship's changed?" Sam wonders if he's been avoiding this truth to dodge pain or self-doubt.

Step 4: Solicit Honest Feedback. Identify trusted people who are willing to tell you the truth, even when it's hard to hear. Ask for candid perspectives on specific behaviors ("What do you see that I might not?") and genuinely listen.

Scenario Example: At lunch the following day, Sam mentions his worries to a mutual friend, Taylor: "I keep thinking Jose is just busy, but it feels like he doesn't want to hang out anymore. Am I imagining things?" Taylor replies gently, "Honestly, I've noticed it too. Since he started his new job, he seems distant from a lot of us—not just you."

Step 5: Intentionally Seek Disconfirming Evidence. When making important decisions or reviewing beliefs, intentionally ask: "What evidence would contradict my view? How might I be wrong?" Consider opposing arguments openly.

Scenario Example: Rather than just accepting his initial narrative, Sam recalls: "When have I seen Jose seem happy to be with me lately?" He remembers a recent game night when Jose made an effort to chat and laugh—maybe it's not total avoidance, but a shift in how often they connect.

Step 6: Notice and Embrace Discomfort. Pay attention to moments when you feel defensive, anxious, or resistant—these may signal the presence of unwelcome truths. Instead of shutting down, explore the discomfort: "What is this reaction trying to protect?"

Scenario Example: Sam realizes he feels both sad and a bit rejected but lets himself sit with those feelings. Instead of immediately rationalizing them away, he writes in his journal: "Maybe our friendship isn't as close now, and that's painful. But pretending it isn't different won't make it go back to what it was."

Step 7: Engage in Structured Self-reflection and Dialogue. Consider therapy, coaching, or participation in group dialogue groups where difficult topics can be explored openly and safely.

Scenario Example: Sam decides to have an open conversation with Jose. He says, "I've felt like things are a bit off between us and I miss how we used to hang out—have I done something to upset you, or is life just really busy?" Jose replies, "Sorry, I've been totally overwhelmed with my new job and family stuff. I haven't been very present; it's not you. Honestly, I've missed you too but haven't reached out."

Step 8: Review and Revise Regularly. Periodically revisit your assumptions, feedback, and past decisions. Make a habit of updating beliefs and acknowledging mistakes as part of growth, not failure.

Scenario Example: Sam reflects on what he's learned:

- He was self-deceiving by telling himself "nothing's wrong" and blaming only Jose.
- Some of the change is outside either person's control.

- Moving forward, Sam decides to be honest about his feelings with friends and not automatically ignore early signs when relationships shift, even if it feels uncomfortable.

What This Shows: Sam's honest, stepwise engagement with his doubts allowed him to face an emotional truth, seek clarification, and adapt. By noticing his self-deception, reflecting, seeking feedback, and embracing discomfort, Sam was able to move from denial to understanding and take healthier, more authentic action—demonstrating that self-awareness is an ongoing practice, not a destination.

Overcoming self-deception is a lifelong process, not a final destination. By practicing self-inquiry, humility, and openness to feedback, individuals can gradually lift the veil, see themselves more clearly, and make wiser, more authentic choices.

Chapter 13
Bits and Pieces

When the world is in bits and pieces,
Each field its own story releases.
We *Whysly* connect,
And strive to reflect
'Til the puzzle of knowledge increases.

There is so much more research and depth of understanding that has emerged which is relative to our focus of *why we can't see what is right in front of us*. Some of the fields less engaged in our previous conversations still have significant contributions to our area of focus. For example, **Artificial Intelligence and Robotics** focuses on developing models and systems that can mimic or respond to human patterns of attention (e.g., visual attention in machine vision, attention networks in deep learning). **Clinical Psychology and Psychiatry** investigates attention-related disorders (ADHD, PTSD, anxiety) and the effects of trauma, mood, and medications on attentional control, and examines therapeutic interventions (like mindfulness) to improve or retrain attention.

Education and Learning Sciences studies how attention affects learning outcomes, classroom engagement, memory encoding, and instructional design. For example, the impact of attention spans, distraction, and multitasking on study and retention. **Ergonomics/Human Factors Engineering** looks at attention in safety-critical environments (aviation, medicine, driving), studying attentional lapses, oversight, vigilance, and interface design for minimizing error. **Human-Computer Interaction (HCI)/Design** explores attention in safety-critical environments (aviation, medicine, driving), studying attentional lapses, oversight, vigilance, and interface design for minimizing error.

Linguistics and Psycholinguistics examine how we direct attention during language processing, reading, and conversation—including phenomena like "garden-path sentences" where our attention is momentarily misled. **Marketing and Advertising** analyze how brands and media compete for human attention, employing psychological techniques to capture and retain focus. Attention is treated as currency: what ads are "seen," noticed, or ignored? **Mindfulness and Contemplative Studies** combine insights from psychology, neuroscience, and spirituality, investigating practices designed to heighten, widen, or stabilize attention. And **Neuroscience** dives deeper into cellular/molecular mechanisms and brain network dynamics underlying various forms of attention. Emerging research explores neural oscillations, neurotransmitters (like dopamine and acetylcholine), and brain connectivity patterns linked to focused and divided attention.

As we can see, there's a larger need for consilience understanding, looking from various perspectives across fields and studies, then we could have imagined. And no doubt we have missed some perspectives, Yet I think we have to pat ourselves on the shoulder for exploring our focal topic so extensively. Recognizing the inability to be exhaustive, some of the pieces mentioned in earlier chapters—theories, approaches, studies, etc.—deserve a bit more attention to ensure understanding their contribution to our focus, as well as a few others we have not yet mentioned. Of course, because of the importance of this topic to our decision-making, and even our everyday living, it touches so many areas of our thoughts, feelings, and actions that a single book could not possibly cover the full scope of this research. Nonetheless, let's use this chapter to briefly explore further those concepts lightly touched, and perhaps to emerge other concepts not yet forwarded which are worthy of reflection.

In this regard and as a future resource reference, we offer brief treatments of related concepts organized in terms of the field from which they emerge. Note this is not claimed to be an exhaustive list.

Anthropology: Cross-Cultural Comparisons, Ethnographic Studies

Cognitive Science: Habitual Perception, Heuristics, Mental Models, Multisensory Perception, and Schema Theory

Communications Studies: Selective Exposure, Spiral of Silence, and Cultivation Theory

Economics: Utility Theory

Education: Critical Pedagogy and Inquiry-Based Learning

Neuroscience: Active Avoidance, Brain Hemispheric Function, Cognitive Load Theory, the Default Mode Network, Neural Plasticity, and Top-Down versus Bottom-Up Processing

Philosophy: The Aesthetic Experience, Deontological Ethics, Ethical Theory (in general), Pragmatism [*Note: Both Deontological Ethics and Utility Theory are a part of Ethical Theory*]

Political Science: Agenda Setting, Framing Theory, Ideology, and Political Socialization

Psychology: Narrative Psychology

Sociology: Hedgemony, Power Dynamics, Social Constructionism, Social Identity Theory, Structural Inequality, and Symbolic Interactionism

Spirituality: Illusion of Separation, Shadow Self, Spiritual Awakening, and Spiritual Ignorance

We begin.

Anthropology

This section includes: Cross-Cultural Comparisons and Ethnographic Studies.

Cross-Cultural Comparisons. These are research methods in anthropology used to identify similarities and differences across cultures, helping to understand the influence of culture on human behavior and social structures. For example, comparing marriage rituals across different societies to understand cultural variations in familial structures.

Through her fieldwork in Samoa, Margaret Mead (1928) examined the lives of adolescent girls, comparing their experiences with those in the United States. She concluded that cultural factors, rather than biological ones, play a significant role in influencing behavior and personality development. Mead's work challenged preconceived Western notions about adolescence as a universally tumultuous period, highlighting how cultural contexts can shape individual experiences in drastically different ways. This study was pivotal in demonstrating the value of cross-cultural research in understanding the diversity of human experiences.

Cross-cultural comparisons are crucial for understanding why we might overlook what is evident in different cultural settings. By contrasting various cultures, anthropologists can uncover underlying assumptions and biases that might not be evident within a single cultural framework. Such comparisons offer insights into how people from different backgrounds perceive and prioritize aspects of their environment. This approach can reveal how cultural lenses shape perception, allowing us to see beyond our immediate biases and become aware of alternative ways of interpreting the world. Thus, cross-cultural comparisons help elucidate the elements of other societies that might initially escape our notice due to our cultural conditioning.

Scenario Example: Consider a cross-cultural study comparing parenting styles in Scandinavian countries with those in Southeast Asia. In Scandinavian cultures, a strong emphasis on independence and outdoor play is prevalent, with children encouraged to explore and learn through self-direction. In contrast, some Southeast Asian cultures might prioritize communal activities and structured learning, focusing on familial bonds and respect for hierarchy. An observer from one culture might initially overlook the benefits and strengths of the other's practices. However, by engaging in cross-cultural comparisons, the researcher can appreciate the different values and outcomes each approach fosters—revealing that what might be ignored or undervalued in one context becomes clear and insightful when viewed through a broader lens. This scenario demonstrates how cross-cultural comparisons can illuminate unseen aspects of other societies and challenge our preconceived notions.

Ethnographic Studies. Ethnographic studies are qualitative research methods used in anthropology to study cultures and communities through direct observation, participation, and interaction over extended periods. For example, an anthropologist living in a remote village for several months to understand local customs and social dynamics.

Geertz's (1973) research emphasized the concept of "thick description" as a way to understand cultures more deeply, suggesting that researchers should not only note observable behaviors but also interpret the underly meanings and symbols that make those behaviors significant within the cultural context. His approach shifted anthropology towards understanding societies by considering intricate layers of meaning thereby enriching our understanding of human cultural expression and interaction.

Ethnographic studies offer valuable insights into why we sometimes fail to perceive what is directly before us. By immersing themselves in other cultures, ethnographers uncover unspoken assumptions, context-specific meanings, and social norms that might be completely invisible to outsiders. Such studies highlight that cultural blind spots are not necessarily about literal visibility but about understanding the invisible structures shaping perception. They reveal how one's cultural framework can either illuminate or obscure elements in a given environment, impacting what is noticed or ignored. Thus, ethnography provides a lens through which to consider how culture influences perceptual awareness.

Scenario Example: Imagine an ethnographer conducting a study in the bustling markets of Marrakech. At first, they are overwhelmed by the colors, sounds, and movement. Over time, through careful observation and interaction, they start noticing subtle social cues and hierarchies among vendors, such as the unspoken seniority-based deference given to particular sellers or the nuanced interactions between buyers and sellers that indicate mutual respect and trust. While these social dynamics are "invisible" at a glance, the ethnographer's gradual immersion allows them to perceive the foundational elements that shape this vibrant scene. This scenario exemplifies how ethnographic study peels back cultural layers, illustrating why we sometimes fail to notice the less overt, yet significant, aspects of what is before us due to our initial cultural preconceptions or lack of contextual immersion.

Cognitive Science

This section includes: Habitual Perception, Heuristics, Mental Models, Multisensory Perception, and Schema Theory.

Habitual Perception. Habitual perception refers to the tendency of individuals to see and interpret the world around them in ways that are familiar and routine. It is shaped by past experiences, learned behaviors, and cultural influences, leading people to perceive their environment in predictable ways. For example, habitual perception can lead you to overlook your own keys on the table because you're so accustomed to placing them there that your brain filters them out as uninteresting.

Because habitual perception is based on repeated exposure to certain stimuli or environments, it can lead to the automatic interpretation of sensory information without active, conscious thought. This has the advantage of enabling efficient processing of information by allowing people to quickly make sense of their surroundings based on past knowledge. However, it can also lead to cognitive biases, where individuals may overlook novel or important details because they perceive situations in a stereotyped manner.

Gibson (1979) introduces the idea of "affordances", which are the possibilities for action that objects and environments offer based on their properties. Gibson argues that perception is not a passive process, but an active one where individuals interact with their surroundings in a meaningful way. Perception is something rooted in a person's habitual interactions with their environment, effectively linked to action in a real-world context.

Related to our focus, this concept is important in understanding how people can sometimes miss changes or nuances in familiar settings, as their brains tend to "fill in the blanks" based on what they expect to see or experience. It highlights the balance between efficiency and flexibility in human perception.

Scenario Example: Imagine a seasoned driver who, after years of experience, instinctively knows how much space is needed to parallel park their car. This driver relies on habitual perception; their brain automatically interprets the spatial environment and distances based on past experiences without conscious deliberation. One day, the driver rents a larger SUV. Despite the different size, they find themselves initially misjudging parking maneuvers. This scenario highlights how habitual perception, shaped by repeated experiences with a smaller vehicle, may not immediately adapt to the nuances of a different-sized car, illustrating both the strengths and limitations of habitual perception in new contexts.

Heuristics. Introduced in the opening of Chapter 10, biases, heuristics are mental shortcuts or rules of thumb that help us make quick judgments and decisions when faced with complex or uncertain situations. While heuristics can be highly efficient, they also predispose us to errors and biases that can obscure what is objectively present. In other words, we sometimes rely on automatic or oversimplified thought processes that lead us to overlook details and nuances—all of which might be "right in front of us."

Tversky and Kahneman (1974) introduce how heuristics guide human judgment in uncertain contexts. While these mental shortcuts help with swift decision-making, they also systematically create as many problems (biases) as they solve—causing individuals to misjudge situations even when details are readily available.

There are several contributions to our focus. First, the idea of *simplification at a cost*. Heuristics reduce cognitive load by focusing on limited cues, often causing us to neglect or dismiss important information that does not align with the heuristic. Second, related to *confirmation and anchoring biases* (see Chapter 10). Common heuristics—like confirmation bias (seeking evidence that supports existing beliefs) or anchoring (over-reliance on initial information)—result in tunnel vision, preventing us from noticing conflicting data. Third, regarding *overconfidence in quick judgment*. Because heuristics can feel intuitively correct, we might overestimate our accuracy, ignoring clear signals that our initial rapid assessment has missed critical evidence.

Scenario Example: In a behavioral economics class, students play an investment simulation game. One student, Linda, anchors heavily on her first successful stock choice, continuing to invest in the same company even though current data—clearly displayed on her screen—shows declining returns. Relying on a heuristic that "this stock was good before," Linda fails to recognize new evidence suggesting it is no longer a wise investment. Despite the objective data being right in front of her in real time, her reliance on an initial mental shortcut prevents her from acknowledging the negative trend until it's too late.

Mental Models. Mental models—which entered our discussion in Chapter 4, The Power of "Why" and again in Chapter 6, Conscious Experience—are internal representations that individuals construct to understand and interpret complex phenomena. They operate as simplified mental simulations—of relationships, processes, and problems—helping people reason about how things work or predict outcomes. However, because they are necessarily incomplete and filtered through personal experience, mental models can lead us to overlook evidence that contradicts our assumptions, leaving us blind to what might be obvious in the environment.

Johnson-Laird (1983) explains how people build and manipulate mental models to reason, make decisions, and predict events. He outlines how mental models, while useful for simplifying reality, can also mislead when they are inaccurate or incomplete, causing individuals to miss evidence that is otherwise observable.

This work relates to our core focus in numerous ways. First, through its connection with simplification and omission. Mental models cut through complexity by focusing on selected features, inadvertently leaving out details that might be crucial or directly in front of us. Second, regarding resistance to revision. Once a mental model "makes sense," it can become entrenched; new or conflicting information that is clearly visible may be dismissed if it does not fit neatly. Third, relative to predictive focus. Mental models favor predicting outcomes in line with priori understanding. Contradictory evidence—even if we encounter it direction—may be absorbed into the model incorrectly or entirely overlooked.

Scenario Example: During a design workshop, a group of novice product developers relies on their mental model that carefully measured parts will always fit together neatly. As they assemble prototypes, they repeatedly overlook a small misalignment—despite physically seeing the parts not lining up. Their collective assumption that "everything should fit if the numbers are correct" leads them to ignore the discrepancy until a mentor draws explicit attention to it, illustrating how deeply held mental models can keep even visible flaws hidden in plain sight.

Multisensory Perception. Related to the fields of cognitive science, neuroscience, and psychology, multisensory perceptions is how the brain processes and integrates sensory inputs to form a comprehensive understanding of the world. Specifically, it is the process by which information from different sensory modalities, such as sight, sound, touch, taste, and smell, is integrated to create a coherent and unified perception of the environment. It is crucial for navigating complex environments and facilitates interactions with the world by enhancing the reliability and richness of sensory information. For example, taste is often enhanced by smell, which is why food can seem less flavorful when one has a blocked nose.

Calvert, Spence, and Stein (2004) delve into topics such as cross-modal interactions, neural mechanisms of multisensory processing, and the impact of multisensory integration on behavior. Their experiments and studies have advanced our understanding of how sensory information is combined in the brain, offering insights into both typical and atypical sensory processing.

Multisensory perception contributes to understanding why we might not perceive what is immediately before us by illustrating the complex processes involved in sensory integration. When senses are working in harmony, the brain synthesizes inputs to form a clear, cohesive picture of the environment. However, when there is discordance, sensory overload, or competing stimuli, the brain might prioritize certain inputs, leading to the exclusion or diminished processing of others. This selective attention can cause us to miss critical environmental cues, demonstrating how perceptual systems are both robust and vulnerable to breakdowns in integration. In scenarios where sensory modalities do not align, important details might be overlooked, despite being present.

Scenario Example: Imagine walking through a crowded city street while listening to music through headphones. Your auditory input is dominated by the music, causing a potential reduction in the perception of surrounding sounds, like an approaching bicycle or someone calling out. Your vision might be similarly focused on finding a specific store, causing less attention to subtle movements in your peripheral vision. This scenario demonstrates how multisensory perception operates—auditory and visual focus on particular stimuli can obscure other sensory inputs, potentially leading to oversights despite your brain's efforts to integrate all available sensory information. Understanding multisensory perception thus sheds light on how attention mechanisms can lead to missing environmental information that is right in front of us.

Schema Theory. Schema theory suggests that our minds rely on cognitive frameworks—known as schemas—to help us quickly interpret and organize new information. These mental structures are formed from past experiences, cultural influences, and repeated exposure to certain patterns or concepts. While schemas offer efficiency, they can also lead to blind spots when incoming information does not match our existing mental templates. In other words, we can fail to see what is literally in front of us if our schemas do not readily accommodate—or flat-out contradict—that new data.

Bartlett (1932) introduced the idea that memory is reconstructive rather than reproductive. This means that people rationalize and reshape new information so it better fits within existing mental structures, underlining how schemas can cause individuals to "miss" or alter what is right in front of them.

There are several connections to our current focus. First, regarding *selective attention* (introduced in Chapter 7). Schemas guide our focus. If we do not have a schema that highlights particular details, we are likely to ignore or overlook them. Second, regarding *confirmation bias* (introduced in Chapter 10). When new information appears that aligns with our existing schemas, we embrace it readily. Data that contradicts our schemas may be dismissed or distorted to maintain our preconceptions. Third, regarding *misinterpretation*. Even when we do notice something, our schemas can lead us to misinterpret what we see, adjusting it to fit established mental frameworks instead of updating our view.

Scenario Example: In a psychology lab, students are shown a series of ambiguous images. Some students, primed with descriptions of animals in previous tasks, quickly interpret the shapes as creatures, while others come up with inanimate objects. Despite viewing the same ambiguous shape, each group's schemas—formed from what they had encountered that day—cause them to "see" different things. Even when told that alternative views are possible, many students struggle to perceive anything outside of their initial schema, demonstrating how these pre-existing frameworks can keep us blind to certain possibilities right in front of us.

Communication Studies

There are three influential concepts from communications theory that shed light on why people sometimes fail to perceive or acknowledge what is plainly in front of them. Each speaks to the subtle (and often unconscious) ways that media environments and social communication can shape, distort, or selectively filter our view of reality. These are selective exposure, the spiral of silence, and cultivation theory.

Selective Exposure. This is the tendency to seek out information that confirms pre-existing beliefs while avoiding content that challenges them. By filtering what we consume—often unintentionally—we reinforce our own worldview and fail to notice or appreciate evidence that contradicts it (Stroud, 2011).

Relevant to our focus, when people are regularly exposed only to content that matches their viewpoints, conflicting information that might be directly observable is overlooked, trivialized, or dismissed. Thus, selective exposure contributes significantly to blind spots and echo chambers.

Scenario Example: Imagine a person who strongly supports environmental regulations and primarily follows news sources and social media pages that advocate for climate action. When encountering credible but opposing information—such as reports highlighting flaws in certain environmental policies—this individual scrolls past or dismisses these stories. As a result, they overlook legitimate concerns and data, reinforcing their belief that there are no reasonable objections to stricter regulations. Even when faced with conflicting evidence in everyday life (such as conversations with informed skeptics), they might tune out or trivialize these viewpoints, remaining unaware of crucial details.

Spiral of Silence. Proposed by Elisabeth Noelle-Neumann (1974), the spiral of silence theory posits that people may refrain from voicing opinions they perceive as unpopular for fear of social isolation. Consequently, these opinions (and relevant information) remain hidden or unspoken.

Relevant to our focus, if certain views or facts are labeled as unacceptable or unpopular, they rarely surface publicly—even when they are objectively valid. The result is a collective blindness to realities that are right there but go unacknowledged due to social pressures.

Scenario Example: Consider a workplace where the majority of employees vocally support a new company policy. A few workers quietly harbor concerns about its effectiveness but, noticing the prevailing sentiment and fearing negative judgment, choose not to express their doubts during meetings. Over time, because hardly anyone voices any opposition, it appears that everyone is on board. Important critiques and suggestions are left unsaid, leading the group to collectively miss problems with the policy that, though obvious to some, are not acknowledged due to pressure for consensus

Cultivation Theory. Developed by Gerbner and Gross (1976), cultivation theory explores how long-term immersion in television (and later, other media) shapes viewers' conceptions of social reality. Repeated media narratives can implant perceptions and biases that distort or overshadow direct personal experience.

Relevant to our focus, when media representations become the baseline for "normal," individuals may be less attuned to real-world indicators that contradict those portrayals. Over time, these internalized media images can overshadow immediate evidence or personal observation, causing people to "not see" what is actually occurring around them.

Scenario Example: Envision someone who watches several hours of crime dramas every week. The shows consistently portray urban living as extremely dangerous, with frequent violent incidents. Even though this person lives in a statistically safe city and has never experienced or witnessed crime personally, they begin to overestimate their risk and avoid certain neighborhoods or activities unnecessarily. Their perception of reality—how dangerous their city actually is—becomes shaped more by fictional media portrayals than by real-world evidence, leading them to overlook the community's actual safety and vibrancy.

Economics

This section includes Utility Theory.

Utility Theory (Ethical Theory). Utility theory, at its core, studies how individuals make decisions to maximize subjective well-being or "utility." Historically derived from ideas in economics and philosophy (e.g., from Jeremy Bentham, John Stuart Mill, and later formalized by John von Neumann and Oskar Morgenstern), utility theory posits that humans weigh costs and benefits to choose the option that provides them the greatest net gain. However, these mental "cost-benefit" calculations can shape our perception in ways that cause us to overlook what might otherwise be obvious. Focused on their theory of games and economic behavior, Von Neumann and Morgenstern (1944) illustrate how individuals (or players) might deliberately ignore certain moves or information if they judge that the effort to consider them outweighs the potential benefits.

There are multiple connections to our core focus. First, in terms of *prioritization of utility*. In seeking to maximize personal benefit, individuals naturally focus on aspects of a situation that promise the greatest payoff. As a result, smaller yet crucial details can go unnoticed because they don't appear to offer significant immediate gain or seem peripheral to current goals.

Second, in terms of *bounded rationality*. Real-world decision-making rarely occurs under ideal circumstances. People have limited time, attention, and cognitive resources, so they often filter out information deemed less valuable—even if it's clearly visible and potentially important later.

Third, in terms of *rational ignorance*. In some contexts, it can be rational for people to avoid the cost of gathering extra information if they see little expected benefit from it. This mindset can inadvertently block awareness of critical data that does lie "in plain sight" but seems unworthy of attention under a utility-maximizing calculus.

Scenario Example: Consider a shopper who is strictly focused on finding the best deal for a specific product. The shopper zeroes in on sale prices for that item, ignoring a prominent in-store announcement about a coupon that could further reduce the bill. In this moment, the well-intended "tunnel vision" on maximizing utility for one product blinds them to an obvious discount that could have improved their overall benefit. By the time they realize the missed opportunity, the cost of backtracking (either mentally or physically) may exceed the perceived gain, reinforcing their unnoticed oversight.

Education

This section includes: Critical Pedagogy and Inquiry-Based Learning.

Critical Pedagogy. This is a teaching philosophy that focuses on unveiling and challenging the power structures, social injustices, and cultural assumptions embedded in educational contexts. Influenced by scholars like Paulo Freire and Henry Giroux, critical pedagogy encourages teachers and students to become co-investigators of knowledge—questioning the status quo and recognizing how cultural norms or hidden biases might distort or obscure one's view of reality. Essentially, the "I can't see what's right in front of me" phenomenon can be traced to institutional and social pressures that often keep certain truths invisible or unspoken. By actively examining these pressures and biases, learners are empowered to reckon with, and eventually see through, misleading assumptions and power dynamics.

Freire (2000) advocates for a problem-posing educational model in which the teacher-student hierarchy is replaced by dialogue and collaboration. By making students active agents in their own learning, Freire proposes a path to critically analyze and transform oppressive structures—encouraging learners to see what they had not previously noticed.

This work contributes to understanding our core theme in several ways. First, through *unmasking ideologies*. Critical pedagogy encourages learners to examine how dominant ideologies shape our perceptions, sometimes blinding us to immediate realities. Second, through *fostering transformative awareness*. It goes beyond surface learning, prompting reflection on personal and collective experiences. Students learn to ask, "Whose perspective is missing here?" Third, through *empowerment through dialogue*. Through open dialogues about power, voice, and representation, students are empowered to notice and openly challenge that which was "hidden in plain sight."

Scenario Example: Mr. Phillips introduces a lesson on the school's dress code, asking his high school English class to analyze how power and control manifest in daily routines. At first, students express superficial awareness: "It's just rules; we have to wear certain clothes." But as they discuss who decided these rules, what cultural or economic factors might be influencing them, and why certain apparel is deemed "inappropriate," they begin to see subtle biases and constraints. In investigating the rationale behind what they had taken for granted, a lively debate emerges—opening their eyes to social and institutional norms that were "in front of them" all along, but remained unquestioned until they engaged in a critically reflective dialogue.

Inquiry-Based Learning. This is an educational approach that places the learner's curiosity, critical thinking, and problem-solving at the center of instruction. Rather than simply presenting factual information for memorization, teachers using inquiry-based methods pose guiding questions, present intriguing challenges, and encourage learners to discover and reason out answers themselves. This active engagement can illuminate blind spots—those "right in front of me" insights that often go unnoticed—by prompting students to ask new questions, see patterns, and investigate assumptions they might otherwise overlook.

Justice and colleagues (2009) explore how inquiry-based learning supports academic skill development by emphasizing exploration and critical thinking. Through real-world examples, the authors demonstrate that crafting thoughtful questions and allowing students to investigate them helps nurture problem-solving capabilities and deeper engagement with content.

Inquiry-based learning makes multiple contributions to understanding our core theme. First, by *encouraging deeper observation*. By consistently asking students "What do you notice?" or "What do you see that might be a clue?", inquiry-based learning trains them to step back and systematically look for details that may otherwise remain hidden in plain sight.

Second, through *promoting ownership of discovery*. Students are more invested when they feel accountable for unearthing answers. That investment makes them more likely to step outside their habitual ways of looking and to notice what they might typically skip over.

Third, by *fostering cognitive flexibility*. Sometimes, "I can't see what's right in front of me" happens because we are locked into one perspective. Inquiry-based learning encourages exploring multiple perspectives and solutions, helping learners recognize when they are stuck and try new angles of thinking.

Scenario Example: Imagine Ms. Reyes's sixth-grade science class. She places a beaker of water on the table—a simple, everyday object. She challenges her students with a question: "How is this water behaving at a microscopic level?" At first, they stare blankly at the beaker, convinced there is nothing special to see. After some encouragement, they start asking each other questions: "How does temperature affect how the molecules move?" "What makes water evaporate?" "Is there gas dissolving in there we can't see?" As these questions pour out, they realize their collective starting assumption—"It's just water"—masked the fascinating world hidden in plain sight. They gather thermometers, use a microscope to examine drops, and record changes in volume over time, unveiling details right in front of them that they originally overlooked. Through inquiry-based exploration, they not only learn about water's properties, but they also understand the importance of asking questions to see beyond the obvious.

Neuroscience

This section includes: Active Avoidance, Brain Hemispheric Function, Cognitive Load Theory, the Default Mode Network, Neural Plasticity, and Top-Down versus Bottom-Up Processing.

Active Avoidance. Active avoidance is a type of behavior where an organism deliberately takes action to prevent encountering an unpleasant or aversive stimulus. This behavior is a learned response that involves recognizing cues predictive of a negative outcome and implementing strategies to avoid it. In humans and animals, active avoidance can serve as a mechanism for coping with potential threats, allowing for adaptive responses in threatening environments. For example, a person quickly changes their walking route after noticing a large, barking dog down the usual path.

From a neuroscience perspective, active avoidance involves several brain regions, primarily the amygdala, the prefrontal cortex, and the striatum. The amygdala is critical in processing threats and emotional reactions, while the prefrontal cortex is involved in decision-making and cognitive control that guides behavior. The striatum plays a role in action selection and reinforcement learning, helping to develop and reinforce avoidance strategies. These areas work together to process environmental cues, evaluate potential dangers, and execute actions to avert negative consequences based on past experiences.

LeDoux's (1996) research explores the neural mechanisms underlying emotions, particularly fear and anxiety, which are central to understanding active avoidance behavior. This work has significantly contributed to the understanding of how emotional and cognitive processes interplay in the brain to guide behavioral responses in threatening situations.

In relation to our core focus, active avoidance can impact our ability to see what's right in front of us because it involves a psychological and neural mechanism where individuals instinctively steer away from potential threats or discomfort, sometimes without rational analysis. This means that when we are conditioned to avoid certain cues or situations, we may overlook opportunities for growth or miss accurate reinterpretations of perceived threats, thus blinding us to the reality of situations that are right before us.

Scenario Example: Imagine an employee who had a past negative experience with public speaking during a team meeting. Whenever there's an opportunity to present ideas, the neural circuits associated with fear and anxiety trigger active avoidance, leading them to decline such opportunities. They focus so much on avoiding the perceived threat of embarrassment that they fail to recognize the supportive and constructive nature of their team now, who are eager to hear new ideas and provide encouragement. This avoidance behavior blinds them to the immediate reality that the current team environment is actually safe and supportive, preventing them from seeing the potential benefits and growth available in the situation right in front of them.

Brain Hemispheric Function. This refers to the specialized roles and ways of processing information typically associated with the left and right hemispheres of the brain. While both hemispheres are deeply interconnected and work together, certain cognitive styles are often linked to one hemisphere more than the other. For example, the left hemisphere is commonly associated with detailed, analytic, and language-based processing, whereas the right hemisphere is often described as more holistic, contextual, and spatially oriented. These differences in perspective, when overemphasized or imbalanced, can sometimes cause us to miss aspects of reality that don't fit our habitual mode of processing.

McGilchrist (2009) provides an in-depth exploration of how each hemisphere contributes unique perspectives and how Western society's historical emphasis on certain analytical, left-hemisphere modes of thinking may lead to an underappreciation of contextual and holistic processing.

Relative to our core focus, when the left hemisphere is dominant in a specific task or context—fixated on the details or a linear narrative—individuals may overlook broader context or nuance. Conversely, if right-hemisphere processes are less engaged, one might lose the "big picture" capacity that could reveal an obvious pattern or significance right before their eyes. Essentially, an overreliance on one hemisphere's style of perceiving the world can create blind spots, because the "other half" of our cognitive framework—more globally oriented or more analytically oriented—isn't sufficiently engaged.

Scenario Example: Imagine a product designer who is deeply skilled in logical analysis—mapping precise specifications, calculating cost ratios, and accurately predicting manufacturing details. Yet, in focusing on these technicalities, they fail to notice the obvious fact that their new design does not match user preferences for simpler, more intuitive interfaces. The "big picture"—what potential customers actually want—

remains hidden in plain sight. This imbalance could reflect an overemphasis on left-hemisphere detail orientation without engaging the right-hemisphere's broader, integrative perspective, thus illustrating how hemispheric differences can lead us to miss what's right in front of us.

Cognitive Load Theory. Cognitive load theory, formulated primarily by John Sweller, examines how the limitations of working memory affect our capacity to process information. According to the theory, individuals have a finite amount of mental resources available at any given time, so when those resources are over-taxed by too much information or too many complex tasks, overall comprehension and retention can suffer. This limitation can lead us to miss details or obvious cues in our environment—even when they're right in front of us—because all our cognitive "bandwidth" is already occupied.

Sweller (1988) forwards how problem complexity and presentation can saturate working memory, clarifying why seemingly simple or obvious details can go unnoticed when the brain is preoccupied with heavier cognitive tasks. Referring to limited working memory, this is directly pertinent to our focus. When working memory is overloaded (e.g., juggling too many tasks or bombarded by complex data), we have less mental space to notice and process new cues that may be easily observable under lighter load.

In terms of intrinsic versus extraneous load, complex material (intrinsic load) is challenging by nature, but external factors like poorly structured information (extraneous load) can further strain cognitive resources. If this strain becomes too great, basic observations are overlooked. When considering the human potential for distraction and overwhelm, high cognitive load can leave us inattentive to relevant details, causing mental "blind spots." Essentially, the brain prioritizes coping with the overload, leaving little capacity for peripheral or seemingly minor but crucial information.

Scenario Example: Cognitive load can leave us inattentive to relevant details, causing mental "blind spots." Essentially, the brain prioritizes coping with the overload, leaving little capacity for peripheral or seemingly minor but crucial information.

Default Mode Network. The "default mode network" (DMN) is a set of interconnected brain regions that tends to become active when we're not focused on a specific external task—often during mind-wandering, daydreaming, or self-referential thinking. While this network serves important functions (e.g., consolidating memories, planning for the future, processing our sense of self), its activity can also compete with our ability to pay attention to the external world in real time. Bucker, Andrews-Hanna and Schacter (2008) explore the biological underpinnings of the DMN, revealing how our brains cycle between external focus and internally oriented activity—and how an imbalance can shape what we do (and don't) notice.

This adds a deeper understanding to our focus. Overactivity of the DMN can keep us preoccupied with internal narratives—rethinking past events, rehearsing future scenarios, or ruminating about ourselves—thus reducing the cognitive resources available for scrutinizing what is happening before our eyes. If habitual mind-wandering remains unchecked, we may fail to register crucial details or immediate cues in our environment. This inattentiveness can compound over time, reinforcing a tendency to "miss the obvious" because our mental spotlight is turned inward rather than engaged with present, external reality.

Scenario Example: Think of someone who routinely drives on "autopilot." While commuting, they might find themselves so lost in thought—reliving a tense conversation or imagining a future vacation—that they barely notice the passing landmarks, changing scenery, or even minor road hazards. When they finally "snap out of it," they realize they've missed not only key external details (like a new billboard or a friend's house along the route) but also the immediate experience of the drive itself. Their DMN-dominated reverie effectively blocks out the full reality right in front of them.

Neural Plasticity. This is the brain's remarkable capacity to reorganize its structure, functions, and connections in response to experiences, learning, or even injury. It shapes how we process information and form habitual patterns of thought or perception. On the positive side, this flexibility enables us to adapt and learn new skills. However, when certain neural pathways become deeply entrenched—through repeated use or reinforcement—they can also limit our perceptions, so that our brains continually "default" to familiar patterns and fail to register alternative perspectives right in front of us.

Doidge's (2007) research explores how the brain's adaptive ability can both empower and confine us, depending on which neuronal pathways become strengthened. It illustrates how habits of mind can shape (and sometimes limit) what we perceive.

There are numerous ways this research supports our understanding of why we can't see what is right in front of us. First, in terms of *habitual pathways*, if your brain is trained to interpret certain stimuli in a particular way—say, you habitually expect negative outcomes—you may miss cues that contradict this outlook. Second, in terms of *reinforcement loops*, the more you rely on the same mental models or narratives, the stronger those neuronal connections become, making it harder to notice new information that doesn't fit. Third, in terms of *filtering mechanisms*, neural plasticity doesn't only help build skills; it also filters experiences through repeated neural circuits. Over time, your brain's "filter" may overlook vital cues or details simply because they aren't part of the existing framework.

Scenario Example: Imagine someone who has practiced piano with poor posture for years. Their neural circuits have become adept at playing in a specific, ingrained way. Even when a new teacher tries to show them a more ergonomic hand position (right in front of them), they struggle to adopt it—automatically reverting to their old habit. This "blind spot" around posture stems from well-established neurocircuitry that makes any alternative feel unfamiliar or wrong. Overcoming this perceptual and physical blind spot requires conscious effort to "rewire" the already-engrained pathways—an apt microcosm of how entrenched brain patterns can obscure what otherwise might be obvious.

Top-Down versus Bottom-Up Processing. Neuroscience distinguishes between these processes. Top-down processing involves the use of pre-existing knowledge, expectations, and experiences to interpret sensory information, while bottom-up processing is driven by the sensory input that builds up to a perceptual experience. For example, when reading a word with missing letters, top-down processing allows you to "fill in" the gaps based on context and past knowledge.

Eysenck and Keane (2015), an in-depth analysis of cognitive processes, explores how top-down processes like expectations and prior knowledge can shape and sometimes distort sensory interpretations, while bottom-up processes allow for raw sensory data to inform cognitive understanding and decision-making. Both have the potential of hindering accurate perception of the environment.

The dichotomy between top-down and bottom-up processing is directly relevant to understanding why we sometimes fail to see what's right in front of us—a phenomenon known as inattentional blindness. (See Chapter 9.) This occurs when the brain is too focused on specific tasks or distractions, heavily relying on top-down processes that filter out seemingly irrelevant sensory stimuli. For instance, when driving while intently thinking about a meeting, a driver may fail to notice a pedestrian crossing the road. Here, the brain's expectation-based top-down processing overshadows incoming sensory information that would normally be processed through bottom-up channels. This can lead to the inability to notice obvious objects, underscoring the power and potential limitations of our cognitive frameworks when they diverge from our immediate sensory environment.

Scenario Example: Imagine you are in a busy airport, trying to find your flight gate while also thinking about an important presentation you're due to give at your destination. As you navigate the bustling terminal, your mind is preoccupied with rehearsing your presentation. Your top-down processes are primed for recognizing and focusing on signs related to your gate rather than absorbing your surroundings. Consequently, you walk past a fellow traveler who is noticeably struggling to carry heavy luggage, failing to offer assistance. Here, your attention is so consumed by the internal dialogue and task at hand that your sensory systems don't alert you to the out-of-the-ordinary scene nearby. This scenario illustrates how top-down processing prioritization can sometimes filter out critical but unexpected sensory information, demonstrating why we occasionally miss what's directly in front of us.

Philosophy

This section includes: The Aesthetic Experience, Deontological Ethics, Ethical Theory (in general), and Pragmatism [*Note: Both Deontological Ethics and Utility Theory are a part of Ethical Theory*]

The Aesthetic Experience. The aesthetic experience refers to a heightened, immersive engagement with art or beauty, which evokes emotional and intellectual responses and appreciation. For example, experiencing awe and introspection while standing before a landscape painting.

The aesthetic experience is fundamental to the human condition because it enriches our lives by fostering emotional depth, intellectual engagement, and a profound connection to the world. Engaging with art and beauty allows individuals to transcend the ordinary, stimulating reflection, inspiration, and a sense of wonder. It promotes empathy by helping us see through others' perspectives and encourages creativity by offering new ways of thinking and feeling. Additionally, the aesthetic experience provides a sanctuary for mindfulness and presence, counteracting the pressures of daily life and enabling moments of peace and contemplation. In connecting people with their emotions and surroundings, it enhances psychological well-being and life satisfaction, making the aesthetic experience a vital component of a fulfilling human existence.

Dufrenne (1973) emphasizes the complex interplay between perception, feeling, and cognition. He argues that the aesthetic experience involves an active dialogue between the observer and the object, revealing layers of meaning and understanding beyond the immediate sensory input. This work is instrumental in articulating how aesthetic experiences transcend observation, facilitating profound emotional and intellectual responses that enrich human understanding.

The aesthetic experience connects to our theme as it highlights how entrenched mental schemas and distracted awareness can prevent genuine aesthetic engagement. Often, the presence of beauty or art is overlooked due to preoccupied minds or habitual perceptions, hindering the ability to fully experience the aesthetic depth that is immediately available.

Scenario Example: Consider John, who walks through the park daily on his commute. Immersed in thoughts about his busy day, he passes by an intricately carved sculpture, barely noticing its presence. One day, prompted by a friend's comment, he pauses to truly observe the artwork. Gradually, he sees the detailed craftsmanship and symbolism in the sculpture, experiencing a rush of emotions and insights he had previously missed. This scenario illustrates how our preoccupations can blind us to the aesthetic richness surrounding us, causing us to overlook the profound experiences lying dormant in the familiar sights of our daily lives.

Deontological Ethics. Most famously associated with Immanuel Kant, deontological ethics centers on the principle that certain moral rules or duties hold universally—independent of outcomes. The rightness or wrongness of an action is determined by whether it adheres to these moral laws (e.g., "keep your promises," "do not lie"), rather than by the results the action produces. Kant's (1785) writings lay out moral imperatives that must be followed categorically—meaning under all conditions—often leading to a strong focus on universal moral principles that can overshadow other contextual truths.

In more recent research related to the field of Knowledge Management, we now recognize that values—to include moral laws—are knowledge and as such are situation-dependent and context sensitive (Bennet & Turner, 2023). In other words, in different situations, one "moral law" may outweigh another "moral law," requiring an understanding from the decision-maker of the implications of actions in a specific context and situation. Apparently, at least at the human level of thought, there are no absolutes.

Let's explore how this can cause us to miss what is obvious or important right in front of us. First, through *rigid rule application*. Deontology emphasizes the inviolability of certain moral duties. Someone strictly following such a rule (e.g., "always tell the truth") may become preoccupied with fulfilling the duty itself and fail to notice context that could qualify or reshape how that duty should be applied. This rigid focus on a single moral dimension creates a form of tunnel vision, obscuring broader situational insights.

Second, through *duty-based blind spots*. Because deontological thinkers often prioritize moral consistency—"it's either always right or always wrong"—they might overlook crucial situational factors (like risk of harm, emotional nuance, or specific stakeholder needs). Such factors can be "in plain sight," but don't register if the guiding question simply is, "Am I following the rule?"

Third, through *moral absolutism*. Deontological frameworks are sometimes interpreted in absolute terms: certain actions are intrinsically right or wrong, regardless of consequences. This absolutism can lead to ignoring real-world complexities—despite their being visible—because deviations might seem to undermine the perceived moral law, introducing a type of "ethical blind spot."

Scenario Example: Imagine an office manager who strictly follows a "no favoritism" policy, believing it's morally imperative to treat every employee identically. A team member asks for extra time off due to a serious family emergency, a fact clearly evident to everyone else. Because the manager is so fixated on upholding an equal policy, they may miss the moral importance of compassion in this specific case—despite the employee's struggle being right in front of them. By adhering rigidly to an overarching rule, the manager overlooks the plain and pressing ethical realities of the situation.

Ethical Theories (in general). Ethical theories—whether deontological (Kantian), utilitarian, virtue-based (Aristotelian), or otherwise—provide frameworks for evaluating right and wrong. While these systems guide moral judgments, they can also create "ethical blind spots" when we become so entrenched in a particular framework or principle that we overlook morally pertinent facts right in front of us.

There are multiple effects of ethics on our core focus. First, in terms of *rule-fixation* (deontology, which is detailed above). A strict adherence to moral rules (e.g., "always tell the truth") can, at times, blind us to the ethical nuances of a situation. We are so focused on the rule that clear contextual information—like another person's vulnerability—barely registers.

Second, in terms of *outcome obsession* (utilitarianism, which is detailed below). If our main concern is maximizing overall well-being or minimizing harm, we might fail to notice immediate injustices that don't obviously affect utility calculations. The push for "the greatest good" can overshadow individual concerns in full view.

Third, in terms of *misaligned virtue* (Aristotelian or Virtue Ethics). Striving for certain virtues (e.g., courage, honesty) can become counterproductive if we ignore how these virtues must be contextually balanced. An overemphasis on being "honest" in one dimension might blind us to the present need for compassion or tact.

Fourth, in terms of *moral certainty*. Individuals may become rigidly self-assured in their ethical stance, forming a bias that filters out contradictory or challenging real-world details. We might not see a harm or a moral obligation staring us in the face because it conflicts with an established moral narrative about who we or others "should" be.

Scenario Example: Imagine a manager who enforces policies with rigid exactness, believing it's the morally righteous path. Employees facing genuine crises (a family emergency or mental health challenge) might be denied necessary flexibility because the manager sees only "unbreakable rules," not the lived context before their eyes. Their moral commitment to fairness, interpreted narrowly, results in missing the immediate moral call to compassion. Here, a well-intentioned ethical principle inadvertently causes them to remain blind to what is plainly happening in front of them.

Pragmatism. Pragmatism, a philosophical movement associated with thinkers like Charles Sanders Peirce, William James, and John Dewey, focuses on the practical consequences of our ideas and beliefs. It asserts that the value or "truth" of a concept emerges from how it works in practice rather than adhering to some abstract or absolute notion of correctness. In other words, what we notice—and how we interpret it—is shaped by our goals, interests, and the tangible outcomes we expect. James (1907) specifically discusses how truth emerges from the practical "cash-value" of ideas.

Meaning, if a proposition or observation doesn't demonstrate use or relevance, we tend to disregard it.

There are multiple facets of pragmatism that are relative to our focus. First, *purpose-driven perception*. From a pragmatist perspective, humans tend to attend to phenomena that appear relevant to their immediate aims or problems. If something doesn't seem practically useful or doesn't fit neatly into our current frameworks, we might overlook it entirely—even when it's in plain sight.

Second, *instrumental knowledge*. Pragmatists such as Dewey (1939) argue that ideas and perceptions function as tools ("instruments") for navigating and solving real-world problems. Ideas and perceptions are tools for practical action. What doesn't serve a clear instrumental function can become "invisible" to us.

Third, *habit and belief-fixation*. Peirce's (1877) notion of "fixation of belief" suggests that once we settle into certain patterns of thinking or acting, we resist contradictory evidence. We may fail to see information that doesn't reinforce our established habits because doing so doesn't serve our working conception of reality in a productive way.

Scenario Example: Imagine a small business owner who is deeply invested in marketing a product in a certain way. Because their primary goal is to attract customers, they focus on promotional strategies—ad campaigns, social media presence, special offers. Meanwhile, an obvious design flaw in the product itself goes unnoticed, despite repeated user complaints, because the owner doesn't perceive fixing it as an immediate solution to driving sales. The pragmatic "end" they're pursuing (generating buzz) overshadows straightforward signals about a fundamental product issue, illustrating how one's practical aims can obscure what is clearly "right there."

Political Science

This section includes: Agenda Setting, Framing Theory, Ideology, and Political Socialization

Agenda Setting. This is a central concept in political science and mass communication that explains how the media and other influential actors determine which issues receive public attention. By focusing on certain topics and underplaying others, the media effectively shapes the perceived importance of social, economic, and political matters. This selective emphasis can cause us to miss what is "right in front of us" when critical problems are not highlighted, or when coverage is consistently framed in ways that make some realities harder to see or believe.

Research by McCombs and Shaw (1972) demonstrates how media coverage influences public perception of what issues matter most. This foundational study maps out how the media doesn't necessarily tell us *what to think*, but significantly shapes *what we think about*.

Contributions to understanding our theme are multiple. First, in *gatekeeping of information*. Editors, news producers, and other gatekeepers decide which stories to promote, effectively filtering the public's field of vision. This can obscure key issues that might be right at hand but are not spotlighted. Second, in *perceived importance*. The more an issue is covered, the more we perceive it as vital or urgent. Issues receiving minimal attention may remain out of sight, leading people to overlook pressing problems simply because they are not presented as such. Third, in *reinforcing blind spots*. Agenda setting interacts with our existing biases. When our preferred media outlets consistently frame or omit certain angles, we are less likely to see them, even if the facts are in plain view.

Scenario Example: In Professor Gómez's public policy class, students are presented with two major national issues: a public health concern (like rising diabetes rates) and a political scandal. Many students struggle to recall any recent coverage on the public health statistics, even though the data are widespread and relatively easy to access. Instead, they focus on the scandal because it has dominated the news cycle for weeks. Through discussion, they realize that agenda setting by the media made the scandal more salient in their minds, causing them to overlook the public health crisis that was "right in front of them" in everyday life. By observing this phenomenon, the students come to grasp how their perceptions are shaped by what the media chooses to spotlight—and how easily unrecognized issues can remain hidden in plain sight.

Framing Theory. In more generic terms, the concepts of "framing" and "frames" are used in Chapter 2, Choice and Agency. The framing theory in political science emphasizes that *how an issue is presented*—its "frame"—can shape people's understanding, interpretations, and actions. Even if the same facts are available, different frames highlight or downplay certain facets of a situation, influencing what individuals consider most salient. Because frames can direct attention to specific values or consequences, they can render certain information invisible or seemingly irrelevant, leading people to "not see" what's right in front of them if it does not fit into the prevailing frame.

Entman's (1993) research clarifies how framing works to select certain aspects of reality and imbue them with greater prominence. By highlighting specific data and contextual cues, frames inform how audiences process and interpret events, showing how some information remains unacknowledged or undervalued.

It has multiple impacts related to our core theme. First, in terms of *highlighting and obscuring*. By drawing attention to specific aspects of a topic, a frame can effectively eclipse other pertinent details, leaving them unseen or disregarded. Second, in terms of *shaping perceptions of relevance*. When a political or media frame defines an issue in terms of "economic cost" or "moral values," it steers audience members toward viewing data through that lens. Information outside that lens—though present—might go unrecognized. Third, in terms of *resonance with audience values*. Frames resonate strongly if they align with an audience's pre-existing beliefs, which can reinforce selective perception and make dissenting or contradictory evidence nearly invisible.

Scenario Example: In a graduate seminar on public opinion, Professor Braga asks students to analyze media coverage of a proposed tax reform. Half the class reviews articles framing the reform as a "necessary fiscal responsibility," stressing how it reduces debt and appeals to middle-class voters. The other half studies coverage framing the same reform as "an unfair burden," emphasizing the negative impacts on lower-wage earners. Though both groups receive much of the same underlying data, each group walks away with a completely different sense of urgency and moral relevance. When they compare notes, they discover the frames they encountered made them more attentive to certain facts while inadvertently glossing over others—exemplifying how our capacity to see what's in front of us can be shaped powerfully by the frame presented.

Ideology. The term "ideology" is used periodically throughout this book. Ideology refers to the system of beliefs, values, and ideas that underpins the political, social, and economic stances people hold. These belief systems often work as interpretive "lenses" through which individuals perceive the world. By reinforcing certain core assumptions, ideologies can cause people to overlook or dismiss perspectives and facts that do not align with their overarching worldview, contributing to an inability to recognize what is "right in front of them."

In his research, Althusser (1971) explores how institutions (like schools, churches, and media) perpetuate ideologies that shape how individuals see and understand their social reality. He illustrates how these "state apparatuses" normalize the dominant ideology, often rendering alternative viewpoints and truths invisible.

Contributions to our theme are multiple. First, related to the *foundation of perception*. Ideological commitments can steer individuals toward seeing only the evidence that supports their worldview, filtering out or distorting other signals. Second, related to *resistance to contradiction*. Because ideology often forms part of one's identity, contradictory information can feel threatening—leading to denial or rationalization of facts in plain sight. Third, related to *social reinforcement*. Ideological communities create environments that echo a shared worldview, making it more difficult to spot issues or realities that fall outside that consensus.

Scenario Example: In Dr. Karim's comparative politics class, students examine different dimensions of nationalism. Students from a strongly nationalist background interpret a global trade agreement as a threat to their nation's sovereignty, focusing on risks of foreign interference. Those from a more cosmopolitan background see primarily the benefits of economic integration and downplay possible risks to local industries. During a structured debate, each side struggles at first to comprehend why the other cannot "simply see" clear evidence of either the concerns or the benefits. Gradually, they recognize that their entrenched ideological positions have led them to selectively perceive information, demonstrating how ideology can obscure what is plainly observable to others.

Political Socialization. Political socialization is the process through which individuals acquire and develop the values, beliefs, attitudes, and knowledge that contribute to their understanding and participation in political systems. This process occurs through various agents, such as family, schools, media, peers, and significant life events, shaping one's political identity and perception over time. The subtlety of this lifelong socialization can lead to individuals aligning with long-standing biases or ideologies, which may inhibit their ability to see alternative viewpoints or recognize changes in the political landscape. For example, a person raised in a politically conservative household may unconsciously internalize conservative values, making it difficult to objectively evaluate liberal policies or ideologies.

A strong reference in this field is the work of David Easton and Robert D. Hess (1962), who significantly contributed to the understanding of early political socialization in their research and publication, *The Child's Political World*. Their studies highlight how the political environment and familial influences can mold children's political attitudes from a very young age. They argue that children develop a sense of political identity and loyalty through daily interactions and observations, often solidifying into long-held beliefs as they mature. This work emphasizes the power of foundational political socialization in shaping future political perception and behavior.

Political socialization is deeply connected to the theme of why we can't see what is right in front of us because it essentially demonstrates how early and ongoing exposure to specific political ideologies can blind individuals to other perspectives or the evolution of societal norms. The ingrained nature of these beliefs means that people may overlook or even dismiss information that contradicts their long-standing views, not out of willful ignorance but due to cognitive bias established through socialization. This habitual adherence to familiar ideologies can prevent individuals from embracing new or different political ideas, despite evidence or changes occurring around them.

Scenario Example: Imagine a young woman named Sarah, who grew up in a household where her family prioritized and revered traditional values and conservative politics. Throughout her upbringing, her parents actively discussed political matters around the dinner table, always reinforcing their preference for conservative viewpoints and skepticism towards progressive ideas. As Sarah progresses through college, she encounters peers and professors who present liberal perspectives and question traditional norms. However, despite the exposure to these new ideas and discussions, Sarah often finds herself unable to critically evaluate or accept liberal viewpoints. Her ingrained political beliefs, shaped by years of family socialization, create a mental block, causing her to frequently overlook or underestimate valid criticisms of conservative policies. This illustrates how Sarah's political socialization makes her "unsee" or disregard the merit of ideas that exist right in front of her, thereby highlighting the challenge of overcoming deep-seated biases formed through early socialization.

Psychology

This section includes Narrative Psychology.

Narrative Psychology. The term "narratives" is used periodically throughout the book. Narrative psychology is the study of *how human beings construct stories to make sense of their experiences and the world around them.* This branch of psychology asserts that individuals frame their realities and identities through the narratives they create, influencing their perceptions, cognitions, and interactions. In this book we have discussed the power of narratives in various ways, as personal and collective narratives, as cultural and moral narratives, and as critical in creating our reality.

While being situated primarily in psychology, the narrative draws heavily from literary studies and communications studies in examining storytelling and its structure. It also intersects with sociology and cultural studies in exploring how cultural narratives shape individual and collective identities.

Singer and Salovey (1993) posit that individuals' self-concepts are tightly woven with autobiographical memories and emotions, emphasizing how past experiences narrated as coherent storylines help shape a person's identity and overall psychological functioning. This approach demonstrates that our remembered stories not only color our present perceptions and expectations of the future but also underpin our ability to act in the world.

The concept of narrative psychology fits seamlessly into the idea that sometimes we can't see what's right in front of us because our perception is filtered through the stories we have constructed. Instead of perceiving reality as it objectively is (if there is such a thing and objectivity), we see it through a lens of our narratives, which can both guide and blind us.

Scenario Example: Imagine Sarah, a successful marketing executive, who constantly narrates her life story emphasizing independence and self-reliance, rooted in a turbulent childhood. At work, she's noticed for her strong leadership and decision-making skills. However, when someone in her team reaches out for help or tries to collaborate closely, she often dismisses these gestures without realizing it. Her narrative of self-sufficiency dominates her perception, blinding her to the benefits of teamwork and support, even when there are potential opportunities right in front of her for growth and improved relations. Despite this, Sarah's entrenched personal story keeps influencing her behavior, illustrating how deeply these narratives can affect not only how we see the world but also how we interact with it.

Sociology

This section includes: Hedgemony, Power Dynamics, Social Constructionism, Social Identity Theory, Structural Inequality, and Symbolic Interactionism.

Hedgemony. In sociology—especially in the work of Antonio Gramsci (1971)—"hegemony" refers to the subtle and pervasive influence that dominant groups wield over social norms, beliefs, and values. It's not always overt or forceful; rather, a hegemonic power often works by shaping what a society considers "common sense." Because of this, certain perspectives, questions, or realities can feel invisible or be dismissed before they're even consciously considered.

How this ties into "not seeing what's right in front of you" is that people's perceptions can be heavily guided by the cultural default. If we grow up, for instance, immersed in specific assumptions about how the world works, we can become blind to alternatives—our sight is filtered through these hegemonic lenses. Even if there are signs all around that things aren't as we've been taught, it can be tough to shift perspective and truly see them. Thus, hegemony plays a large part in shaping how we interpret our surroundings, influencing what appears obvious, what remains hidden, and even how we decide what is worth noticing

Scenario Example: Imagine a public school system in which the history curriculum mainly highlights the accomplishments of one dominant cultural group. Textbooks show that group's literature, art, scientific achievements, and political milestones as the central pillars of national identity. Students from other cultural or ethnic backgrounds see little or no representation of their heritage or contributions—yet, because this is the standard curriculum, everyone may view it as "common sense" or "naturally important." They grow up simply accepting these textbooks as truthful and comprehensive.

In this scenario, the dominant group's norms and perspectives are assumed to be universal, exerting hegemonic power. People can have difficulty "seeing" what's missing—namely, that the curriculum overlooks or marginalizes other groups—because their entire upbringing reflects and reinforces that part of society as central and the rest as peripheral. This illustrates how hegemony can obscure alternative views and experiences, making it hard to notice or challenge what's right in front of us.

Power Dynamics. This refers to the ways in which power is distributed, exercised, and negotiated in social relationships. In sociology, "power" isn't just about overt authority (like a boss telling an employee what to do). It also includes subtler, often invisible ways of influencing what people believe, how they act, and what they consider normal or possible. These dynamics can be seen at every level, from interpersonal relationships (e.g., within families) to larger social and political structures. Lukes (2005) identifies different "dimensions" of power, including the capacity to shape people's preferences and perceptions—often so subtly that power dynamics become hard to see or challenge. This perspective shows how power can hide in the most automatic aspects of social life.

In relation to our focus, because power can operate in indirect ways—shaping cultural norms or unspoken rules—a society's power structures can cause us to overlook certain inequalities or assumptions. When dominant groups hold power, they can frame what counts as "common sense," making opposing viewpoints or realities less visible. Essentially, those with less power may be conditioned to "go along" with the prevailing norms and may struggle to identify or name the very forces that constrain them.

Scenario Example: Imagine a corporate office where a team manager uses informal tactics to influence behavior. Instead of explicitly ordering staff around, the manager frequently drops hints about what ideas are "compatible" with company culture. Employees who suggest approaches that do not align with these hints find their ideas quietly sidelined or ignored. Over time, the team members learn to self-censor, offering only the types of proposals that fit the "unspoken rules." Even without direct confrontation, the manager's power has shaped the group consensus, limiting what gets proposed and pushing team members toward a uniform viewpoint—ultimately hiding the fact that many workable ideas never even make it to the table.

Social Constructionism. Social constructionism is a theory that suggests that our understanding and knowledge of the world are not inherent or natural but are instead constructed through social interactions and collective agreements. For example, the concept of money holds value because society collectively agrees to its value, despite it being paper or digital entries.

Berger and Luckmann (1966) argue that societies create realities through shared understandings and communications. They explore how everyday interactions contribute to the construction of knowledge, positing that reality itself is a human product. This work highlights that social institutions and phenomena are created by people interacting together and coming to a common understanding. Berger and Luckmann's treatise laid the groundwork for exploring how societal constructs influence perceptions of truth, identity, and meaning in varying cultural and social contexts.

Social constructionism relates to our inability to see what is right in front of us by explaining how societal frameworks and norms shape perception. Since reality is socially constructed, the filters through which we view the world can obscure elements that fall outside or challenge established constructs. What a culture deems "real" or "important" directs attention, often sidelining what doesn't fit into pre-defined narratives. This means that certain realities may remain invisible until they are acknowledged by society, demonstrating that our perceptions are not just objective sensorial processes but are deeply influenced by social and cultural boundaries.

Scenario Example: Consider a classroom setting where gender roles are being discussed. In this environment, traditional gender norms might be so ingrained that non-binary identities remain invisible to many until they are explicitly recognized and discussed. A student might pass unnoticed when their identity doesn't fit into the binary categories reinforced by the societal constructs present. However, once the class engages in a dialogue about gender as a social construct with various expressions beyond traditional binaries, awareness starts to shift. This scenario illustrates how the social construction of concepts like gender can obscure other realities, highlighting the importance of social discourse in acknowledging and bringing to perceptual foregrounds what was previously unseen.

Social Identity Theory. Social identity theory—introduced in the discussions of implicit bias and ingroup/outgroup bias in Chapter 10—explains how individuals derive part of their self-concept from the social groups they belong to—such as ethnic, religious, professional, or political groups. Developed by Henri Tajfel and John Turner (1979), the theory posits that people categorize themselves into in-groups and out-groups, often favoring their own group to boost self-esteem. This process of in-group/out-group formation can lead to stereotypes, biases, and group-based comparisons shaping our perceptions.

In relation to our focus, Tajfel and Turner's work demonstrates how group membership influences attitudes and behaviors, illuminating the psychological mechanisms—like in-group favoritism and out-group discrimination—that can mask or distort our view of social realities.

Scenario Example: Imagine an office where two departments—Marketing and IT—frequently clash over project priorities. Marketing staff identify strongly with their creative, fast-paced approach and view the IT team as "too rigid," while IT workers consider themselves logical and strategic, labeling Marketing as "impractical." Each department is quick to credit its own successes to "inherent talent" while attributing the other's achievements to luck. When client feedback suggests a blended approach might be best (creative ideas supported by strong technical infrastructure), both sides resist. They don't see the value in the other group's perspective because their social identities and in-group bias overshadow clear evidence (i.e., feedback pointing to the benefits of collaboration). Such group-based perceptions—rooted in social identity—keep them from recognizing what's right in front of them: that combining strengths could yield better results.

Structural Inequality. Structural inequality refers to the systemic and enduring patterns by which certain groups are privileged and others are systematically disadvantaged across societal institutions—such as education, healthcare, housing, and employment. These inequalities aren't just about individual prejudices; they are embedded into the very structures, policies, and practices that govern society. As a result, they can be difficult to see because they often manifest as the "normal" or "default" way of doing things, just like how hegemonic ideas feel like common sense (Tilly, 1999).

Here's how it ties into the challenge of not noticing what's right in front of us:

- *Institutional Blind Spots*: When institutions (e.g., schools, legal systems, workplaces) consistently produce outcomes favoring some groups over others, people within those institutions may come to see these patterns as just the way the system works. It becomes difficult to perceive the structures behind the disparities—whether it's hiring practices or curriculum choices—because they are woven into everyday routines and appear "natural."
- *Reinforcing Hegemonic Norms*: Structural inequality often operates hand-in-hand with cultural hegemony. Dominant norms and values can frame inequalities as products of merit, effort, or individual shortcomings instead of pointing to systemic barriers. Consequently,

societal narratives may place responsibility for unmet needs on marginalized individuals rather than questioning or rethinking the institutional frameworks themselves.

- *Obscuring Alternative Explanations*: Because structural inequality is so deeply embedded, those who benefit (often unknowingly) from certain advantages might struggle to recognize the roles of privilege and exclusion in their success or comfort. Meanwhile, those who experience systematic barriers can be overlooked or dismissed if hegemonic narratives suggest everyone has an equal shot. The result: we fail to see the full picture of how inequalities accumulate and persist.
- *Normalizing Disparities*: Ongoing social inequalities—from wage gaps to disparities in healthcare access—can become so "normal" that they cease to register as problems needing collective attention. This normalization process keeps us from questioning the structures themselves, perpetuating a cycle in which the inequality remains hidden in plain sight.

In short, structural inequality ties directly into the theme of "not seeing what's right in front of us." By cloaking unfair social patterns behind everyday assumptions and the seemingly natural functioning of institutions, structural inequality and hegemonic norms work together to make these realities feel almost invisible—at least until we lean in and consciously unpack them.

Scenario Example: Imagine a small business owner who is deeply invested in marketing a product in a certain way. Because their primary goal is to attract customers, they focus on promotional strategies—ad campaigns, social media presence, special offers. Meanwhile, an obvious design flaw in the product itself goes unnoticed, despite repeated user complaints, because the owner doesn't perceive fixing it as an immediate solution to driving sales. The pragmatic "end" they're pursuing (generating buzz) overshadows straightforward signals about a fundamental product issue, illustrating how one's practical aims can obscure what is clearly "right there.

Symbolic Interactionism. Symbolic interactionism is a sociological framework that examines how individuals create meaning through social interactions, focusing on the use of symbols and language as core elements in these interactions. For example, the act of a handshake can be understood as a symbol of greeting, respect, or agreement depending on the cultural context.

Herbert Blumer (1969)—building on the ideas of his mentor, George Herbert Mead—underscores the importance of symbols and language as mediators of human interaction and meaning-making. He outlines how social reality is constructed through individual and collective interpretations of symbols. Blumer's work emphasizes the dynamic nature of social life, where meanings are continuously negotiated and reshaped through interactions. This perspective has been influential in fields such as sociology, anthropology, and communication by providing insights into how social identities and realities are constructed.

Symbolic interactionism relates to the phenomenon of overlooking what is in front of us by highlighting how interpretation, rather than simple sensory perception, shapes our understanding of reality. In social contexts, symbols and interactions guide attention, often filtering out what is deemed irrelevant based on shared meanings and societal norms. This process can result in unnoticed elements if they do not fit into our existing symbolic frameworks. By focusing on how meaning is negotiated, symbolic interactionism suggests that perceptual blind spots are not purely sensory but are also shaped by social constructs that dictate which symbols and interactions are prioritized.

Scenario Example: Picture a social setting in a diverse workplace where employees are mingling during a company event. A new hire comes from a culture where maintaining prolonged eye contact is considered confrontational. As they interact with colleagues, they avoid direct eye contact, which some coworkers might misinterpret as a lack of confidence or engagement. These coworkers unconsciously focus on nonverbal symbols they are accustomed to, potentially overlooking the new hire's verbal contributions and expertise. This scenario illustrates how symbolic interactionism helps explain why certain social signals are highlighted or ignored, emphasizing the role of culturally informed interpretations in shaping social perception. Understanding these dynamics allows for greater awareness of how societal symbols can influence what we notice or fail to observe in our interactions.

Spirituality

This section includes: Illusion of Separation, Shadow Self, Spiritual Awakening, and Spiritual Ignorance

Illusion of Separation. In many spiritual traditions—particularly those emphasizing non-duality (e.g., Advaita Vedanta, certain Buddhist schools)—the "illusion of separation" refers to the mistaken notion that we exist as completely isolated individuals, fundamentally disconnected from others and from the broader cosmos. This perspective posits that at a deeper level, everything is interconnected, and the sense of being a separate self is an illusion born of limited perception. Hanh (1998) discusses the principle of "interbeing", emphasizing that all phenomena exist in relationship to one another. Recognizing this interconnected reality can dissolve the illusion of isolation, thereby clarifying one's direct experience.

Relative to our focus, when we so deeply believe we are separate from everything else, we may fail to notice the underlying cohesion and mutual influence at play. This illusion can cause us to feel alone or threatened, prompting us to tune out or block what doesn't fit our narrow view of self versus other—a key way we end up "missing" vital insights or possibilities in our immediate surroundings. By clinging to this sense of separateness, we overlook the flow and interdependence that could reveal new perspectives, solutions, or paths forward.

Scenario Example: Consider a person who, feeling perpetually alone, believes that no one truly cares about or understands them. Because of this mindset, they don't notice small acts of kindness—like coworkers offering help or neighbors inviting them to social gatherings. Instead, they interpret these moments as trivial or insincere. In reality, genuine willingness to connect is "right in front of them," but their ingrained belief in utter separateness blinds them to those everyday gestures of support. Only when they start to question the assumption of existing in complete isolation might they begin to see the extent of care and community that has, in fact, always been there.

Shadow Self. As influenced by the work of Carl Jung—and embedded in many spiritual and depth psychology traditions—the "shadow self" refers to the parts of ourselves we tend to repress or deny. These can include impulses, traits, and feelings deemed unacceptable by our conscious self-image or societal norms. Because the shadow remains largely unconscious, it can manifest as a "blind spot," concealing truths about our perceptions, behaviors, and relationships. Jung (1959) delves into the concept of the shadow as part of his broader exploration of archetypes and the individuation process, discussing how hidden aspects of the psyche affect our conscious view of reality.

Relative to our focus, when we are unaware of our own shadow elements, we can project them onto others or overlook them entirely in our day-to-day experiences. In other words, our unacknowledged biases or fears may keep us from recognizing certain insights, signs, or lessons in our environment. If we're unconsciously pushing uncomfortable truths into our shadow, we remain partly "in the dark," unable to fully perceive what's really going on around us or within us

Scenario Example: Imagine a person who frequently finds colleagues "untrustworthy" or "too competitive." They complain of being exploited or undermined at work. However, upon deeper reflection, it might turn out that they themselves harbor a desire for recognition and sometimes manipulate situations to their own advantage—but this aspect of their personality is pushed into the shadow. Because they refuse to see that trait in themselves, they spot it everywhere else, distorting how they perceive and interact with their peers. They fail to notice genuine opportunities for collaboration (right in front of them) because these shadow elements obscure or twist their view of the situation. By recognizing and integrating the shadow side—owning their own competitive streak—they might start noticing more trustworthy alliances and positive dynamics that were always there, just hidden behind their projections.

Spiritual Awakening. Spiritual awakening often describes a shift or expansion in consciousness—moving beyond habitual, ego-driven thought patterns into a more present, direct experience of reality. In many traditions, awakening (or enlightenment) involves realizing that our everyday "ordinary" view of the world is limited or obscured by mental habits, emotional attachments, and the illusion of a separate self. Through this new awareness, we begin to perceive life in a fuller, unfiltered way. Tolle (1999) highlights that most people live with a constant stream of mental noise, preventing them from fully experiencing the present moment. By becoming deeply present and conscious of this mental activity, one can open up to a deeper perception—essentially seeing what's been in front of them all along. Interestingly, this is the concept behind social and cultural awakening (or "woke"), which has become misaligned in today's political environment.

In relation to our core exploration, prior to awakening, one's attention may be scattered by constant mental chatter—ruminating on past regrets or future worries—which prevents noticing the richness of the present moment. The transcendent or enlightened perspective involves recognizing how these thought patterns confine our vision, limiting our capacity to see events, objects, and people clearly as they are now. As awakening unfolds, a person may experience a softening or dissolving of ingrained identities and fears, revealing a wider, more direct awareness that was always available but frequently overlooked.

Scenario Example: Picture someone who walks daily through a beautiful neighborhood park, yet barely registers any sensation of the trees, birdsong, or fresh air. Instead, they're absorbed in concerns about workplace drama or replaying past conversations in their head. A moment of spiritual awakening might occur when, perhaps prompted by a personal crisis or simply by a still moment, they suddenly notice the park's vibrancy and the peace it can offer. In that instant, they perceive a unity with their surroundings—no longer an isolated self lost in worries. That experience can mark the realization that life's beauty, and answers to some struggles, were present the entire time, hidden behind mental preoccupations.

Spiritual Ignorance. The concept of spiritual ignorance refers to an inability (or unwillingness) to recognize the true nature of reality. In many wisdom and religious traditions—particularly those influenced by Buddhist thought—ignorance is framed as "spiritual" when it pertains to fundamental misunderstandings about life's deeper reality or the nature of consciousness. This distinction highlights that we're not talking just about a lack of factual knowledge (like not knowing a piece of trivia) but rather an inability to perceive the underlying truths at the heart of existence:

- It involves overlooking the interdependence and impermanence of all phenomena.
- It includes clinging to a mistaken sense of a permanent, separate "self."
- It manifests as a resulting dissatisfaction or suffering, because we remain bound in illusions that conflict with how things truly are.

The term "spiritual" underscores that this misunderstanding, or *avidyā*, isn't purely cognitive. It's bound up with our very sense of self and our experience of meaning or purpose—in other words, our spiritual dimension. When this core way of seeing is off, it obscures any deeper or more integrated understanding of ourselves and the world, thereby keeping us from fully recognizing "what's right in front of us." This ignorance is seen as the root cause of suffering because it distorts perception of ourselves, others, and the world around us. Rather than acknowledging the impermanent and interconnected nature of life, we cling to illusions about our own separateness and permanence, which obscures what's directly in front of us (The Dalai Lama, 1997).

In relation to our focus, if we operate from a state of spiritual ignorance, we tend to filter our experiences through preconceived notions, attachments, and fears. We might seek permanence in fleeting circumstances or see division where there is actually interdependence. As a result, we remain blind to deeper truths within our everyday experiences—a classic case of being unable to perceive the reality that is right there, waiting to be noticed.

Scenario Example: Consider someone who repeatedly chases after external achievements (e.g., promotions, financial gains, social status) as the sole measure of personal success. They remain dissatisfied, constantly seeking the "next thing" in hopes it will bring lasting happiness. From a spiritual ignorance standpoint, they are failing to notice the ever-present sources of peace and contentment—like relationships, gratitude, or personal growth—right in front of them. These more enduring forms of fulfillment escape their attention because their worldview is narrowly fixed on a mistaken notion: that happiness comes solely from external validation. Overcoming spiritual ignorance, in this sense, means recognizing that what they genuinely need may already be accessible if they look beyond those illusions.

Final Thoughts

As we've moved through the diverse fields engaged across these rich areas of research—with each contributing its own lens and conceptual vocabulary to this additional learning—it has become even more apparent just how profoundly interdisciplinary our understanding of "not seeing what is right in front of us" truly is. Each field, whether cognitive science with its focus on mental shortcuts and models, communication studies with perceptions filtered by media and social dynamics, or neuroscience's insights into brain architecture and plasticity, brings essential pieces to a complex puzzle. What stands out upon reflection is not only the breadth of these perspectives, but also the degree to which they converge around a set of core themes: the limitations of perception, the power of context, the influence of group dynamics, and the capacity for reflection and change.

What this synthesis reveals is that our perceptual blind spots are rarely the result of single factors. Instead, they emerge from the intricate interplay between individual cognition, social environment, cultural influences, and even philosophical worldviews. By juxtaposing theories from education with those from philosophy, politics, or spirituality, and all the fields and studies represented in this book, we see that blind spots are at once deeply personal and fundamentally collective. This realization encourages humility and curiosity in our own judgments, reminding us that insight often arises when we are willing to look beyond the boundaries of our habitual thinking or disciplinary silos.

Moreover, recognizing these patterns offers not only explanation, but also hope. If perceptual limitations are shaped by cognitive habits, media exposure, social pressure, and educational practices, then they can also be addressed—through intentional effort, critical inquiry, open dialogue, and even spiritual growth. The path toward greater awareness is therefore both individual and communal, requiring both self-reflection and collective action. Ultimately, what is at stake is not just accuracy of perception, but the quality of our decision-making, relationships, and participation in society. In recognizing our blind spots, we gain the power to expand our vision—bit by bit, field by field, and moment by moment.

Chapter 14
Augmented "Seeing"

My thoughts were all tangled in glue,
'Til a chatbot said, "Let's think this through!"
We *Whysly* explored,
Asked questions galore,
Now I'm seeing the world fresh and new!

Throughout this book we've explored how and why we fail to see what is right in front of us—psychologically, neurologically, and socially. Why do obvious truths slip past unnoticed, or crucial details vanish from our gaze—even as our eyes, ostensibly, take in the whole scene? The question is as old as human thought, animating disciplines from philosophy to psychology and now, in the age of artificial intelligence, computer science. "Seeing is believing," we like to say, but the deeper truth is that what we see depends as much on our attention, beliefs, intentions, and histories as on the world itself.

Human blindness is not just a matter of sensory limitation, but a byproduct of the mind's remarkable efficiency. As we have learned, we attend selectively, filter aggressively, and reinterpret reality to fit the contours of our experience and expectations. This is the puzzle of the "invisible gorilla," of change blindness, of bias, of cognitive dissonance, and self-deception: phenomena that lay bare the gap between what we observe and what we are able—or willing—to acknowledge.

Now, as machines increasingly model and even extend human cognition, a new lens opens on these ancient questions. Artificial intelligence systems, constructed to "see" and "know," echo our susceptibilities: they too overlook, misjudge, and fall prey to their training and context. Yet precisely where AI fails, it can illuminate the mechanisms of our own cognitive blind spots, suggesting new ways to surface, fill in, or even transcend them.

Drawing parallels with AI recognition systems, in this chapter we ask what our digital mirrors reveal about the architecture of attention, memory, and bias. And as we learn together, we speculate about the coming era of cognitive augmentation: where artificial companions may help us notice what we otherwise miss while posing new challenges about what it means to truly *see*.

The Many Faces of Human Blindness

As we have learned, there are many faces to human blindness. For example, we've explored attentional blind spots, that which we miss without realizing it. Inattentional blindness occurs when attention focused elsewhere renders us "blind" to otherwise salient events. Change blindness describes our failure to notice substantial changes in a visual scene, so long as those changes occur during a visual disruption. But these blind spots are not failures of the sensory organs. Rather, they reveal that *seeing* is a process of *constructive selection*: we perceive what we attend to, and much else is filtered out.

We also explored cognitive dissonance and belief as filters. Moving beyond the raw perceptual sphere, our *worldview*—shaped by experience, culture, identity, and emotion—profoundly conditions perception. Cognitive dissonance describes our tendency to avoid incongruent facts or reinterpret them to preserve a coherent self-understanding. Similarly, self-deception manifests not as an accident, but as an adaptive filter, allowing us to overlook inconvenient truths. In both domains, "not seeing" is less a bug than a feature—a byproduct of brains evolved for *efficient survival* in noisy environments, not for objective truth.

And along the way we learned about schemas, priors, and the predictive mind. Emerging theories such as *predictive processing* or *predictive coding* frame the brain as a "prediction machine" constantly generating hypotheses about the world and updating them with sensory evidence. We "see" what we predict and adjust only when errors or surprises are large enough to demand correction. As a result, perception is always "top-down" as much as "bottom-up"—and our priors can either illuminate or obscure what is in front of us.

Machine Learning as Model and Mirror

Modern AI—especially deep learning—has reached or surpassed human-level performance in narrow recognition tasks (e.g., image classification, face recognition, language translation). These systems, such as convolutional neural networks (Krizhevsky et al., 2012) and transformers (Vaswani et al., 2017), work by abstracting patterns that distinguish different categories based on exposure to massive datasets.

For example, a deep neural network trained to identify cats in images is shown millions of pictures, each labeled "cat" or "not cat." Gradually, by adjusting the strengths of connections between artificial neurons, the system comes to recognize subtle cues: the shape of ears, the spacing of eyes, fur texture, and so on. Over time, it abstracts a general "template" of what a cat typically looks like—not by memorizing specific images, but by capturing recurring features across hundreds of thousands of examples. This is the power of Big Data. When presented with a new photo, the AI analyzes its visual elements and checks if these match the learned patterns. If they do, it classifies the image as a cat; if not, as something else. This ability to detect and classify input by abstracting patterns from large datasets underlies much of AI's current success.

In the realm of language, transformer-based AI models like GPT (Radford et al., 2018) learn to understand and generate text by analyzing vast collections of books, articles, and websites. These systems are not programmed with explicit grammar rules; rather, they ingest millions of sentences and gradually recognize statistical patterns in how words and phrases tend to occur together. For example, when asked to complete the sentence "The cat sat on the ___," the model predicts "mat" because it has seen that word frequently follow that phrase in its training data. In answering questions or translating between languages, the AI draws on these abstracted patterns of word order, grammar, and common usage, enabling it to generate plausible sentences—even for situations it has never encountered before. Just as image recognition models generalize from visual features, language models generalize from linguistic patterns found across massive datasets.

Yet these systems are, like us, products of their histories: their "knowledge" is bound by their training data, architectures, and objective functions. AI systems are notorious for failing under "distributional shift"—that is, when the kinds of data they encounter in real-world deployment differ from the data they were trained on. For example, a model trained to recognize animals in crisp, studio-lit photos may struggle with images taken in the wild, under poor lighting or with unfamiliar backgrounds. In distributional shift, the statistical properties of the input—such as color, context, noise, or even the types of objects present—change, often in subtle ways the model is not prepared for (Ovadia et al., 2019; Geirhos et al., 2020). The model's confident predictions can become unreliable precisely because it has not learned to generalize beyond its narrow training context.

Perhaps most dramatically, AIs can be "fooled" by minute alterations in input (Goodfellow et al., 2015). A slight perturbation to a digital image—imperceptible to humans—can cause a CNN (convolutional neural network) to confidently mislabel a panda as a gibbon. For example, researchers have shown that by adding a barely detectable "noise" pattern to an ordinary photo of a panda, a CNN that would normally recognize the animal with over 99% confidence might instead declare it is looking at a gibbon with equal certainty. This brittleness is not alien to human cognition: it mirrors our own "jumping to conclusions" and our tendency to misclassify or ignore data outside familiar contexts.

The parallels with human cognition are not surprising considering AI is a creation of the human mind. The *ways* in which AI fails are remarkably similar to human errors. For example, both systems (1) rely on prior experience ("training data" or "schemas"); (2) use context to "fill in" ambiguous or missing information; (3) exhibit bias—taken-for-granted assumptions about what is relevant or possible; and (4) are subject to "attention bottlenecks"—they cannot process all input at once and must prioritize. These represent just some of the many parallels researchers have observed; in fact, ongoing studies continue to reveal further unexpected similarities in the ways humans and AI organize, generalize, and sometimes overlook information.

But there are also crucial differences as well. AI systems lack *self-reflection* and *meta-cognitive awareness* of what they do not know, unless so engineered. Humans, in contrast, can (if motivated and with effort) become aware of their own blind spots. Human perception is further shaped by our embodied and multisensory experience of the world, integrating sight, sound, touch, and context in ways that help us adapt flexibly to new or unexpected situations. This robustness allows us to recognize familiar objects in unusual settings—a dog in a cartoon, for example, or a friend's voice on a distorted phone line. By contrast, most AI systems process only one type of input (such as images or text), and are often brittle when faced with unfamiliar conditions or subtle changes in their environment.

Human errors, moreover, can arise from a host of additional factors—attentional lapses, motivation, emotion, or social pressures—that simply do not exist for AI. While AI mistakes are typically the product of statistical quirks in training data or design, human mistakes may be colored by desire, fear, habit, or group dynamics.

Another fundamental difference concerns the *capacity for learning new concepts and transferring knowledge*. Humans are remarkably efficient at grasping new ideas or recognizing entirely new objects from just a handful of examples—a skill known as "few-shot learning." We are also adept at drawing analogies and applying insights learned in one context to solve problems in another, often unrelated, domain.

In contrast, most AI systems require exposure to vast amounts of labeled data to reliably learn a new category or skill, and they typically struggle to generalize knowledge outside their specific training domain. This limitation further highlights the distinct ways in which humans and machines encounter, interpret, and adapt to new information in the world around them. Taken together, these differences highlight that, even when the errors of machines seem to mirror our own, the underlying mechanisms and limitations of human and artificial perception remain worlds apart.

Illuminating the Unknowns of Human Blindness

By explicitly modeling and analyzing where and why AI systems fail to recognize what is "there," researchers have found new ways to think about the strengths and traps of human cognition. For example, systematic error analysis in deep learning has revealed that AI models often engage in "shortcut learning"—relying not on meaningful, robust features, but on superficial correlations that happen to work in training data.

A famous case is image classifiers that distinguish cows from camels not by the animals' inherent features, but by the common backgrounds in the training photos (green grass for cows, sand for camels) (Geirhos et al., 2020). When these cues change—say, a camel on grass or a cow on sand—the AI can fail spectacularly. These findings have inspired parallels in psychology, suggesting humans too may unconsciously use contextual "shortcuts" or surface cues when making judgments, contributing to our blind spots and biases.

Moreover, the field of interpretability in AI (Doshi-Velez & Kim, 2017) attempts to "open the black box," paralleling psychological efforts to make implicit cognitive processes accessible to conscious scrutiny. Techniques such as saliency mapping, for instance, visualize which parts of an image most influence a neural network's decision, sometimes revealing surprising dependencies or overlooked cues. In natural language processing, attention maps can highlight which words carry the most weight in a model's translation or summary, facilitating comparisons with human reading strategies and drawing attention to mismatches.

The table below are elements of that "black box" that has been opened. For each interpretability technique we include an AI example and a human analogy, highlighting how the technique parallels methods or experiences from human cognition or psychology.

Technique	Description	AI Example	Human Analogy
Saliency Mapping	Highlights input features (pixels, words) most influential for a decision.	Visualizing which areas of a photo drive an AI's "cat" classification.	Eye-tracking studies showing which parts of an image people look at first when identifying animals.
Feature Visualization	Generates synthetic inputs that maximally activate certain neurons or layers, revealing what they "look for."	Creating patterns that show what a neuron perceives as "striped fur."	Drawing a picture of a "typical" cat after seeing many photos—an amalgam of features you remember.

Occlusion Sensitivity / Perturbation	Removes or alters input parts to see which regions most affect the output.	Blurring different image patches to see which most influence the AI's choice.	Covering parts of a photo with your hand to see if you can still recognize what's in it.
Activation Maximization	Finds inputs that strongly activate a specific neuron/output.	Generating an image to "max out" the AI's "face detector" neuron.	Asking someone what features make a face most recognizable, then exaggerating those in a drawing.
Concept Activation Vectors (TCAV)	Measures how much a user-defined concept shapes model predictions.	Quantifying if "stripes" or "wheels" are important for certain image classes.	Rating how much seeing "stripes" or "wheels" helps you tell if something is a tiger or a car.
LIME (Local Interpretable Model-agnostic Explanations)	Builds a simple, interpretable model (e.g., linear regression) for a single prediction to explain it.	Explaining why a review is "positive" by ranking the influence of certain words.	Having someone justify their opinion on a movie by pointing to the keywords in their review.
SHAP (Shapley Additive exPlanations}	Having someone justify their opinion on a movie by pointing to the keywords in their review.	Showing which words led a model to flag an email as spam.	Asking people which words in an email made them suspect it was spam, and how much each contributed.
Attention Mapping	Visualizes which input elements a model "pays attention" to when making decisions (common in language models).	Revealing which words in English and French are paired by a translation AI.	Noting which words a translator mentally aligns when converting sentences between languages.

This quest for interpretability echoes psychological methods—like think-aloud protocols or cognitive task analysis—that try to surface the often invisible steps involved in human reasoning. By comparing where both humans and AI succeed and stumble, researchers are increasingly able to illuminate the hidden mechanisms underlying what we do—and do not—see.

AI as Companion, Corrective, and Co-Conspirator

Ultimately, the fact that we cannot "see" what is right in front of us is not, in itself, a sign of failure or folly. It is the mark of minds with finite processing power, evolved for efficiency rather than completeness. Yet, as our world grows more complex and our artificial companions more capable, acknowledging and mitigating our "blinks" becomes both a technical and an ethical imperative.

The challenge is not just to "fill" our gaps, but to surface and reflect on them; not to eradicate bias or blindness, but to cultivate tools—and habits—that make our awareness more porous, agile, and open. As we build and use AI systems to augment perception and recognition, we are not only outsourcing vision—we are, in a sense, beginning to see with new eyes.

Here are some ways that can happen.

Filling the Blinks. Suppose we build tools—AI-based *cognitive prostheses*—capable of detecting gaps in human attention, knowledge, or recognition, and flagging or "filling in" what we're missing. Such tools could operate in real time via augmented reality (AR), wearable sensors, or decision support systems. The biological retina has a literal *blind spot* where the optic nerve exits the eye; the brain "fills in" the missing information based on adjacent signals. Could an *AI companion* serve a similar role for higher-order cognitive blind spots—surfacing unacknowledged data points, unexamined assumptions, or unexplored alternatives? We might call this process Cognitive Blink Completion, Attentional Surfacing, Meta-Attentional Augmentation, or Epistemic Scaffolding. Any of those terms would make sense.

Extended Mind and Epistemic Partnership. Clark and Chalmers's (1998) theory of the *extended mind* suggests that cognition can "leak out" into the world and our tools—like pen and paper, or now, a smartphone or AI assistant—are as much a part of our thinking as the brain itself. From this view, AI-based augmentation could scaffold our awareness, make us less likely to miss the "obvious," and function as a kind of *collective cognition partner.*

Risks, Trade-Offs, and New Kinds of Blindness. Augmentation is not risk-free. Outsourcing our attention and noticing to AI systems might create new dependencies, or *meta-blind spots*—places where the "system" overlooks or distorts reality in ways we cannot easily audit. We risk creating new *filter bubbles* (Pariser, 2011), new "deciders" about what deserves attention. We may also, over time, atrophy our own capacities for vigilance and critical noticing.

There is a further ethical dilemma: Who decides what our prosthetic mind draws our attention to? Can an AI system truly know "what matters" in context? What happens when the priorities of the AI designer diverge from those of the human user?

And, as we move down this road of thought, consider:

- Would you choose to have an AI overlay that points out every perceptual or cognitive "miss"? Would this lead to deeper understanding, or to overload and distrust of your own senses?
- How can we co-design AI companions that *supplement* rather than *supplant* our unique—even necessary—forms of blindness?

Synthesized After Thoughts

Our inability to see what is right in front of us is neither flaw nor trivial quirk; it is a defining feature of human cognition—born of necessity, shaped by evolution, and sharpened by experience. We filter, fill in, and sometimes fabricate our perceptions not out of laziness, but as a survival strategy, conserving mental resources while navigating a world awash in complexity and ambiguity. Yet, these same adaptive mechanisms leave us exposed to error, bias, and self-deception, often blinding us to obvious truths and urgent realities.

In exploring the parallels between the human mind and artificial intelligence, we gain invaluable perspective on our cognitive blind spots. AI's recognition systems falter in ways that mirror—and sometimes magnify—our own selective seeing, its successes and failures exposing the hidden linchpins of context, expectation, and prior knowledge. Just as AI's "misses" teach us about the contours of machine vision, they sharpen our understanding of the architecture and vulnerabilities of our own awareness.

The prospect of AI-augmented perception—systems designed to illuminate what we overlook—offers hope for greater clarity and deeper understanding. Yet, such augmentation carries risks as well as promises: the danger of new dependencies, unexamined values embedded in our tools, and a creeping atrophy of our ability to notice and reflect unaided.

Ultimately, the challenge is not to eliminate blindness altogether, but to cultivate both humility and creativity in the face of our inevitable limitations. By recognizing both the necessity and the cost of our cognitive "blinks" and "blanks", and by thoughtfully integrating artificial partners into our perceptual lives, we can aspire not toward perfect vision, but toward wiser seeing—rooted in awareness of what we miss, and open to the world's ever-unfolding possibilities.

Chapter 15
Revisiting Chapter 1

We stumble past truths every day,
While clear paths hide in the gray.
Our questions fall short,
Of insight and sort,
Yet *Whysly* will show us the way.

For those of you who have reached this chapter, I applaud your persistence, your continued interest in this very real and challenging human issue, and your openness to learning. In the writing, such has been my journey as well.

Let's go back to our opening chapter where we described "A World in Chaos: Economic and Social Turmoil in the U.S." as an example. Now that we have built a foundational understanding of why we can't see what is right in front of us, let's read through this early example of our current world with new eyes and explicitly cross-reference our learning so that we make the interconnections of our integrative/consilience approach transparent. Annotations prompt us to go deeper, expanding our understanding and allowing for moments of personal engagement. By thematically linking narrative to theory and practice, we reinforce the book's structure as a tool for transformational learning.

To the left is the original narrative, broken with spaces between key logical/subject breaks. Some original sentences may be split to clarify where elements shift. To the right is the analysis referencing corresponding chapters, theories, perspectives, and concepts throughout the book.

Text (Excerpt)	Book Elements in Play / Analysis
In a nation where the principles of democracy and equality were once seen as hallmarks of society, the present moment finds many Americans in a state of uncertainty.	**Ch.1 Introduction** Backdrop of turmoil, national identity, and challenging established ideals. [*Explore* Ch. 4—**The Power of Why**. Reflect on the *Why* of the U.S.] **Ch.12 Political Science** (Ideology, Political Socialization; **Ch.12 Sociology** (Power dynamics, Structural Inequality) **Ch.12 Blindness** (Political, Cultural).
Fractures of political divisiveness, deepening economic inequality, and widespread social unrest color the national mood. Recent years have seen a shift toward more extreme rhetoric, with ideological divisions sharpening and eroding a sense of shared reality.	**Backdrop** of turmoil, national identity, and challenging established ideas. **Ch.8 Blindness** (Political, Ideological, Cultural, Perceptual) **Ch.9 Biases** (Confirmation Bias, In-group/Out-group Bias, Status Quo Bias). [*Explore How* these rifts are widened: see **Ch.12 Communications Studies** (Selective Exposure, Spiral of Silence)]

Policy debates increasingly hinge more on entrenched party lines and less on shared facts or nuanced reasoning.	**Ch.12 Political Science** (Agenda Setting, Framing Theory) with pubic discourse shaped by simplification, polarization. **Ch.12 Communications Studies** (Cultivation Theory, Spiral of Silence)
Simultaneously, a series of high-impact decisions are made under the guise of fiscal necessity: essential social programs—Medicaid, Social Security, education subsidies, and public assistance—face deep cuts.	**Ch.2 Agency** (How much say do individuals feel they have?) [see **Ch. 3** (Cognitive Reappraisal) and **Ch.12 Education** (Critical Pedagogy).] **Ch. 8 Blindness** (Moral Blindness, Ideological Blindness) **Ch.9 Biases** (Self-Serving Bias, Projection Bias, Normalcy Bias) **Ch 10 Cognitive Dissonance** (Political Science/Economic rationale vs. values) **Ch.12 Political Science** (Framing Theory—language of "necessity", Ideology) **Ch.12 Philosophy** (Deontological Ethics, Utility Theory, Ethical Theory).
Supporters of these measures point to budget shortfalls and rising national debt as reasons for "belt-tightening." Detractors, meanwhile, argue these are disproportionate sacrifices demanded mainly of the most vulnerable.	**Ch.5 Consilience** Comparison of frames: Ch.12 **Philosophy** (Deontological Ethics) and religious/ethical framing. **Ch.8 Blindness:** (Cultural, Economic, Social). **Ch.9 Biases:** (Omission Bias—ignoring vulnerable groups). **Ch.10 Cognitive Dissonance** (Supporters must reconcile pain to others with fiscal goals; Detractors confronted with lack of agency) **Ch.12 Sociology** (Structural Inequality, Hegemony, Social Identity Theory). **Education:** (Access, Equity).
The rationale of "returning tax benefits" is invoked, even as these tax breaks overwhelmingly favor the highest earners and largest corporations.	**Ch.8 Blindness** (Economic Blindness) **Ch.9 Biases** (Self-Serving Bias, Optimism Bias, Normalcy Bias} [*How the affluent rationalize gains; how the public may fall prey to Normalcy Bias; raises transparency questions.]* **Ch.12 Political Science** (Utility Theory, Ideology)
These moves raise profound ethical questions about the allocation of resources and the foundational values underlying policy.	**Ch.4 Why** (*Ask:* What is our *WHY* as a society? What are our obligations to each other?") [See Appendix A: Socratic Questioning] **Ch.12 Philosophy** (Ethical Theory—Deontological vs. Utilitarian)
As critical services erode, those at the social margins—children, the elderly, low-income workers—face compounding hardships. Rising tariffs and inflation further restrict access to healthcare, education, and basic necessities, intensifying the gulf between privileged and precarious.	**Ch.7 Attention/Intention** (What does society pay attention to, and what slips from view?) **Ch.8 Blindness** (Economic, Cultural, Social, Temporal Blindness—slow erosion is often invisible until severe) **Ch.11 Self-Deception** (Societal Level) **Ch.12: Sociology** (Structural Inequality, Social Identity Theory, Power Dynamics)
Proponents of austerity measures assert their necessity for economic "health." Yet critics document a steady deepening of inequality and loss of opportunity among disadvantaged populations.	**Ch.8 Blindness** (Economic, Cultural) **Ch.10 Cognitive Dissonance** (Competing narratives—efficiency vs. equity) **Ch.11 Self-Deception** (Selective justification, "protection becomes a trap" for those clinging to narratives) **Ch.12 Sociology** (Structural Inequality)
Language itself obscures impact: policy and media discussions frequently focus on "fiscal responsibility," glossing over human consequences and rarely centering the lived experiences of those most affected.	**Ch.8 Blindness** (Perceptual, Ideological, Moral) **Ch.12 Communications Studies** (Framing Theory, Spiral of Silence, Cultivation, Selective Exposure—how language shapes perception) [Framing is a tool of power. See **Ch. 6** Exercise 2: memory vs. reality (explores perceptual limits)]

Despite the justifications, long-term projections suggest these very cuts accelerate federal debt, ignoring the nuanced and interconnected effects, and contradicting stated fiscal objectives.	**Ch.8 Blindness** (Intentional Blindness, Cultural, Psychological) **Ch.9 Biases** (Confirmation, Omission, Normalcy, Projection) **Ch.10 Cognitive Dissonance** (Denial/avoidance of awkward truths) **Ch.12 Political Science** (Agenda Setting) [Here we see the ironic reversal—intentions producing the opposite effect. *Reflect* on *How* complexity is minimized by popular narratives.]
Many citizens—overwhelmed by the chaos or worn down by relentless newsfeeds—remain unaware or actively avoidant of these consequences.	**Ch.6 Conscious Experience** **Ch.7 Attention** (Attentional Blink, Inattentional Blindness) **Ch.11 Self-Deception** (Active avoidance as an individual and collective) **Ch.12 Cognitive Science** (Habitual Perception) **Ch.12 Neuroscience** (Default Mode Network) [Take a moment to reflect. *Ask*, How do YOU decide what to attend to, and what to ignore?]
Misinformation, spin, and deepening partisanship fragment the information environment. Social media algorithms and echo chambers intensify this, pre-selecting the narratives users encounter and reinforcing ideological silos.	**Ch.6 Conscious Experience** (Hawkins' Model and the Modular Approach of Perception) **Ch.8 Blindness** (Ideological) **Ch.9 Bias** (Confirmation Bias) **Ch.12: Communications Studies** (Selective Exposure, Spiral of Silence, Cultivation Theory) [*Reflect* on your own media environment. You might choose the Socratic Questioning approach (see Appendix A).]
Those who speak up against these trends encounter dismissal, sometimes hostility, as concerns are reframed as political attacks or dismissed as exaggeration.	**Ch.9 Status Quo Bias** [Consider the important roles of a whistle-blower, or independent Inspector General.] **Ch.10 Cognitive Dissonance** (Defensive routines, rejection of discomfort) **Ch.12 Political Science** (Agenda Setting, Ideology)
This selective blindness fosters business-as-usual, even as underlying social tensions intensify, threatening the underpinnings of civic trust and cohesion.	**Ch.5: Consilience** (Missed opportunity for integrated perspective) **Ch.8: Blindness** (Moral, Cultural, Perceptual, Inattentional) [*Review Challenge*: Glance through Chapter 8 and explicitly name the potential kinds of blindness at play. *Reflect:* What do you see and not see in your own community?]
Communities now face spiraling healthcare expenses, rising poverty, and the fraying of public infrastructure and safety nets—challenges that demand collective awareness and coordinated response.	**Ch.3 Breaking Through** (Transformational Learning, Dialogical Interpretation, Cognitive Reappraisal) **Ch.14 Augmented Seeing** (AI as model/mirror for illuminating complex patterns) **Ch.15 From Passivity to Agency** [*Consider*: What would collective agency look like here? What needs to happen for the invisible to become visible?]
The critical question emerges: How do ordinary people come to terms with and respond to a reality that is actively denied or minimized by prevailing narratives?	**Ch.2 Choice and Agency** **Ch.3 Breaking Through** (Support During Belief Transition) **Ch.12 Education** (Inquiry-Based Learning) **Ch.12 Spirituality** (Spiritual Awakening) Ch.14 **I/We Create Our Reality** [This can be the moment of awakening. *Reflect*: Is this the reality you are satisfied/happy with?]
Though the magnitude of the challenge may seem overwhelming, it also brings an inflection point: the possibility for communities to re-examine their beliefs, engage critically, and reclaim the guiding ideals of democracy, dignity, and equity.	**Ch.3 Breaking Through** (Transformational Learning) **Ch.4 The Power of "Why"** (Articulating Your Own "Why") **Ch.5 The Convergence of Consilience** (Applying a Consilient Approach to Our Consilient Exploration) **Ch.15: Awakening to Creative Reality-Shaping** (Exercises 1 and 2) **Appendix A: Socratic Questioning**

From Blindness to Creative Agency

By now, if you've traveled with me through narrative, theory, and deep personal questioning, you have carried a considerable burden—emotionally, intellectually, and perhaps even spiritually. If you feel a disquiet at what has been learned, know you are not alone; that discomfort is a sign of awakening, and a necessary prelude to change.

This annotated revisit of our opening story was not only an exercise in understanding, but a map for the journey ahead—one in which each of us holds a pen, a brush, a lens. With the groundwork laid—on how blindness operates, how agency is constrained and reclaimed, how transformation can occur—the question is no longer whether we see, but *what we are willing to do with that new understanding*.

Take a moment to list, or simply imagine, your spheres of influence. These are the spaces where your sight matters most.

- **Personal**: What beliefs or habits will you now question or shift?
- **Family and Friends**: How might new conversations begin? Who can you encourage to see differently?
- **Workplaces and Schools**: Is there an unspoken norm ready for re-examination? How could a process or policy be made more just?
- **Community and Civic Life**: Where do the needs of the "invisible" show up? Where do collective stories need unearthing or telling?

You have acquired frameworks—the vocabulary of blindness, bias, cognitive dissonance, and the possibility of consilience. Yet seeing only matters if it leads to renewed intention and action. Even small acts of attentive seeing—naming an injustice, asking a deeper question, refusing a comforting but blinding narrative—can begin the work of transformation.

Seeing Forward: Practicing New Vision

Throughout this book, we have explored the profound question of why we so often fail to see what lies in plain sight—whether in ourselves, our relationships, or the societies we shape. We've uncovered the many forms of blindness and bias that keep critical realities hidden: from the limits of our attention, to the influence of habit and narrative, to the comfortable but constraining stories we tell ourselves and each other.

But understanding *why* we don't see is only the beginning. The real challenge—and promise—lies in what comes next: *How can we train ourselves to notice, to question, and to act upon what is right before us, but formerly overlooked?*

The simple exercises that follow are designed as practical bridges between insight and action. Each invites you to step beyond habitual perception, challenge your established frames, and practice new ways of seeing—both individually and alongside others. These are not only intellectual tasks, but lived experiments, echoing the central principle of this study: that awareness, agency, and transformation begin with the courage to *notice what is hidden* and to bring it, compassionately and critically, into the light.

Several exercises focus on your personal patterns of attention and omission. Another invites you to explore the blind spots woven into relationships, institutions, and collective narratives. While they are all short, all are meant to help you build the muscles—of perception, reflection, and agency—that make the invisible visible.

As you work through these practices, remember:

- The act of seeing more clearly is ongoing.
- Discomfort is a sign that you are learning to look differently.
- Even small experiments with awareness and action can ripple outward, changing what is possible for individuals and communities alike.

The path toward seeing what is right in front of us is rarely easy—but it is always worthwhile. Below are seven ways to begin this journey. Take your time. Let what emerges surprise you. Return to these exercises as your vision expands, cycles deepen, and new blind spots come into view. And make up more exercises of your own.

Mapping My Spheres of Influence

Purpose: This exercise helps you recognize where you already have power to see differently, to question, and to spark change—in your own life, relationships, and community.

Instructions:

1. Draw three concentric circles on a blank page. Label them:
    - *Personal* (innermost)
    - *Relational* (middle)
    - *Collective* (outermost)
2. In each circle, list areas of your life where you have influence.

- *Personal*: Habits, beliefs, self-talk, attitudes
- *Relational*: Family, friends, colleagues, neighbors
- *Collective*: Civic groups, religious communities, school boards, online forums, voting, workplace culture, advocacy groups

3. For each circle, pick one area where "seeing differently" could make a difference. Next to this entry, jot down one small but concrete step you could take. *Examples*: Ask a new question; name a blind spot; challenge a routine; invite honest feedback; bring a hidden issue into the open; support a marginalized voice.
4. *Reflection*: Looking at your spheres, where is your influence strongest? What one action are you willing to commit to this week?

Blind Spots Journal

This is a seven-day consciousness challenge.

Purpose: Become aware of the many kinds of "blindness" operating in everyday life—perceptual, emotional, cultural, and more. Discover what you've been missing, and how you might begin to see further.

Instructions:

1. Each day for the next week, carry a small notebook or open a digital note just for this purpose.
2. Every day, jot down at least three observations in answer to:
   - "When, today, did I realize I had missed something?"
   - "Where did I catch myself making an assumption, glossing over an issue, or failing to notice something important?"

   Examples: A person whose viewpoint I ignored; an issue I dismissed; a routine I followed without questioning; a moment of discomfort I distracted myself from.
3. At the end of each day, review your notes. For each blind spot, briefly note:
   - What type of blindness was at play? (See Chapter 8.)
   - What made it possible to see it now?
   - What might I do differently next time?

4. At the end of seven days, reflect:

    - What patterns do you see in your blind spots?
    - Where have you started to see more clearly—and how do you feel about it?

Rewrite the Narrative (Narrative Psychology)

Purpose: Practice "narrative agency" by reimagining and retelling a dominant story—shedding light on what is normally invisible, and opening room for justice, complexity, or hope. (See "Narrative Psychology" under Psychology in Chapter 13.)

Instructions:

1. Identify an issue you care about—perhaps one from the stories in this book, or one you encounter in your own community. *Examples:* poverty, public education cuts, polarization, health care access, or climate change.
2. Write down the prevailing narrative as you typically hear or see it:

    - Who is centered?
    - Whose experience, needs, or perspective is left out or minimized?
    - What assumptions does it make?
    - Where might it gloss over suffering, agency, or complexity?

3. Now, apply what you've learned about "blindness," about consilience, and about attention.

    - Who is missing from this story?
    - What humanity, struggle, or possibility remains unseen?

4. Rewrite the narrative, bringing in missing voices and complexity. Imagine you're telling this story to someone who's never heard it—what needs to be included for it to be more complete, honest, or just?
5. (Optional) Share your rewritten narrative with someone else. Notice their reaction.
6. Reflection:

    - How did it feel to "see" and give voice to the unseen?
    - What difference might such a reframing make?

Socratic Questioning on a Current Event

Purpose: Move past passive absorption of news or social debate. Practice critical, Socratic inquiry to reveal unseen frames, assumptions, and interests. (See Appendix A on Socratic Questioning."

Instructions:

1. Select a current news item, political debate, or story that stirs strong feelings.
2. Read (or watch/listen) carefully. Once you feel you "know what's going on," set it aside.
3. Using the Socratic approach from Appendix A, ask yourself:
    - What are the key claims and assumptions in this story?
    - Whose interests does this narrative serve? Who benefits if I (or the public) see it this way?
    - What alternative framings or facts might also be true, but are omitted?
    - Which voices, perspectives, or impacts are missing or minimized?
    - What biases (confirmation, omission, status quo, etc.) could be at work—both in the media and in my own thinking?
    - If I held the opposite view, what evidence or experience would I point to?
4. After reflecting on these questions, notice:
    - Did your sense of certainty or outrage change?
    - Do you wish to seek out counter-voices or more information?
    - What, if any, action or further inquiry does this spark?
5. Journal your insights.

Attend to the Unseen: Mindful Walk

Purpose: Practice focused attention and observation as a metaphor for seeing "the invisible" in your surroundings—cultivating new habits of awareness and compassion.

Instructions:

1. Choose a familiar environment—a street, park, public space, place you regularly pass through.

2. Set a clear intention: "On this walk, I will notice everything I usually overlook."

3. As you walk, slow down. Tune in to your senses. What do you *see* , *hear* , *feel* , and *smell* that you often filter out?

    - Who or what do others ignore?
    - Signs of care (flowers, repair work, greetings) or neglect (litter, isolation, decay)?
    - People on the margins, or gestures of connection or exclusion?

4. If you like, take notes or photos, or even make a simple sketch of what you discovered.

5. After the walk, reflect:

    - What surprised you most?
    - What does this reveal about "blindness," bias, or the limits of your normal attention?
    - What would change if you made this habit of seeing a regular practice?

Collective Awakening Conversation

Purpose: Turn individual insight into dialog and shared discovery. Invite group reflection and agency in challenging local "blind spots."

Instructions:

1. Gather a small group (3–6 people)—family, friends, colleagues, or community members.

2. Invite them into a shared conversation about "what we don't see."

    - Begin by describing your own process or learnings from this book (optional: read a passage or show your "spheres of influence" map).

    - *Ask*: (1) "What is often invisible, ignored, or accepted in our shared space (home, school, workplace, community)?" (2) "Whose needs, stories, or suffering are routinely unseen here?"

3. Use concepts from this book—types of blindness, bias, agency, "why" questions—to deepen the dialogue.

4. As a group, choose one hidden pattern or issue that resonates for further exploration.

- What small action could make the invisible more visible here?
- Who might you need to listen to, invite, or lift up?

5. If possible, make a commitment together: one experiment, conversation, or project (even a single act) to shift awareness.

Follow-up: At a later date, regroup to debrief what changed, and what you learned.

Design Your Own Awakening Ritual

Purpose: Ground your learning and intention in a daily or weekly practice, creating habits that help you see, question, and respond more consciously.

Instructions:

1. Review the exercises above. Choose one (or a combination) that made the biggest impact.
2. Decide how you will integrate this into your routine—as a morning intention, an evening reflection, or a weekly check-in.

   Examples:

   - Ask yourself, "Where am I blind today?" each morning.
   - Do a weekly "blind spots journal" review.
   - Schedule a monthly narrative-rewriting session with a friend or group.
   - Carry the Socratic questions in your wallet or phone for quick reference.

3. Commit to this ritual for 21 days. Put reminders where you'll see them.
4. Journal insights, changes in perception, and any ripple effects you notice.
5. At the end of 21 days, pause to reflect:

   - How has your seeing changed, inwardly and outwardly?
   - What new possibilities or calls to action have emerged for you?

6. Celebrate your growth—and decide what to continue or deepen.

These exercises are short beginnings, not endings. Several longer exercises—building on these—are included in the next chapter. With each small act of questioning, seeing, and choosing, you contribute to a world where more becomes visible. By taking step by step with courage, curiosity and commitment, more is possible to see.

Chapter 16
I/We Create Our Reality

In the tapestry woven with will,
Our beliefs shape the world we fulfill.
As *Whysly* we mold,
New stories unfold
Crafting realities wondrous and still.

Joy, empowerment, and psychological growth come from realizing we actively participate in—indeed, *create*—our own reality. Traditionally, reality is conceived as a fixed landscape onto which humans must fit their perceptions. Yet a rich tradition in philosophy and psychology argues that individuals actively participate in *constructing* their reality through perception, belief, and meaning-making. This chapter explores the profound sense of joy and agency that can arise when one recognizes their role as an active creator of experience, not a passive recipient. The discussion is anchored in constructivist psychology, existential philosophy, and educational theory.

For much of history, reality was seen as an objective given—one either adapted to it or suffered discord. However, research in psychology and philosophy demonstrates that perception is always filtered, shaped, and reconstructed by the mind (Kelly, 1955; Berger & Luckmann, 1966). To realize that one is not just a spectator, but a *co-author* of personal reality, can engender a sense of possibility, playfulness, and profound joy. Let's explore what it means to "create our reality," the sources of joy in this discovery, and its implication for agency and well-being.

Synthesis of Theoretical Foundations

In personal construct psychology, George Kelly (1955) advanced the idea that humans use "personal constructs"—unique, individually-developed systems of interpretation which function like personal hypotheses—to build their realities. There is no vision of the world unmediated by thought; each person improvises on reality's themes, selecting, interpreting, and reinterpreting their circumstances through personal frameworks. Thus, we have creative, dynamic roles in this process.

Today this has been affirmed through neuroscience research as detailed in Hawkins' (2021) Thousand Brains theory. In Chapter 6 on the Conscious

Experience we explored the physiological and cognitive mechanisms of awareness, discovering that awareness is not passive, but actively constructed and highly selective, meaning that our conscious experience is shaped by both biological imperatives and learned models. Our brains work as active sculptors of reality based on our beliefs, context, and goals.

In Chapter 7 on Attention and Intention and Chapter 8 on Relativism, we found that our subjective reality is shaped by what we expect or hope to see, rather than the totality of what is actually present. Thus, when our thoughts and feelings are focused and consistent we are effectively creating the reality of the moment in which we exist based on what captures our attention. In Chapter 10 we further learned that cognitive biases distort one's perception of reality, with an unwillingness to "see" the reality that contradicts our beliefs or experiences. And in Chapter 15, and moving into this chapter, we confirmed that we attend selectively, filter aggressively, and reinterpret reality to fit the contours of our experience and expectations.

Berger and Luckmann (1966) articulated how reality is built not only individually, but socially. They argue that much of what we take for granted as "reality"—from daily routines to social roles and cultural norms—arises from ongoing processes of human interaction. They show that knowledge is socially negotiated: as people share language and interpretive frameworks, they "construct" the world together, generating and sustaining meanings that become so ubiquitous we forget they are human-made. This work is foundational for understanding that our realities are not just discovered, but continually created and sustained by ourselves and those around us. Even apparently "given" realities (like gender norms or work roles) are made, not found.

Existentialist thinkers deepen this, highlighting both the anxiety and exuberance that come from recognizing one's radical freedom—and responsibility—in shaping the meaning of the world. For example, Jean-Paul Sartre's (1943/2003) phenomenological existentialism maintains that humans are "condemned to be free": we always interpret, choose, and create meaning in the absence of predetermined essence or structure. For Sartre, reality is open-ended and projective; each person must decide what their world and actions mean, and this freedom is both an enormous responsibility and a source of exuberant possibility.

Maurice Merleau-Ponty (1945/2012) argues that perception is not an internal mirror of the world, but an engaged, bodily *process of participation*. Our experience of reality is always anchored in our lived perspective, shaped by context, action, habituation, and bodily awareness. This means that the "given" world is always colored by the way we move through and interact with it—underscoring the creative, embodied nature of perception.

The Constructivist View that Beliefs Shape Reality

The powerful role of our beliefs in shaping our reality was introduced in Chapter 2, and social constructionism is discussed in Chapter 13 under "Sociology." With a focus on how we create our reality, we expand on this earlier learning.

The constructivist view that beliefs shape reality is rooted in the philosophical idea that our understanding of the world is constructed through the lens of our individual and collective beliefs, experiences, and social contexts. Constructivists argue that reality is not something that exists independently of our perceptions; rather, it is formed through our interactions, cognitive processes, and societal influences.

The foundational constructivist thinkers include names that you may well recognize. Jean Piaget (1968, 1971), known for his theory of cognitive development, argued that humans construct knowledge through experiences and interactions with the environment. Piaget built on the work of John Dewey and Kurt Lewin, and in turn his experiential learning model was refocused and expanded by David Kolb, J. E. Zull and David Hughes Bennet. Lev Vygotsky (1986) emphasized the social aspects of learning, proposing that community and culture significantly impact cognitive development, and that higher mental functions develop through social interactions. We agree. Today, we recognize that humans are primarily experiential learners, with the expanded model (ICALS—The Intelligent Complex Adaptive Learning System, which is the human) bringing the mode of social learning into the experiential learning model. The five modes of the ICALS model (surrounding "self") include: Concrete Experience, Reflective Observation, Abstract Conceptualization, Active Experimentation and Social Engagement (Bennet, Bennet, & Turner, 2018, 2022).

In the field of education, constructivist approaches stress active learning environments. Students are encouraged to explore, ask questions, and derive solutions through experiences rather than rote memorization. In the field of psychology, in therapeutic contexts, changing an individual's beliefs can be a pathway to altering their perception of reality, hence improving mental well-being. From the viewpoint of science and epistemology, constructivism challenges traditional positivist approaches in science, suggesting that what we accept as scientific knowledge is influenced by historical and cultural contexts, and is thus subject to change as these contexts evolve.

Key features of constructivism include:

- **Subjective Experience**. Constructivism emphasizes that our perception of reality is influenced by personal experiences and interpretations, meaning each individual may construct reality differently.
- **Social Influence**. Social interactions and cultural contexts play a significant role in shaping beliefs, which in turn influence how individuals and groups perceive reality.
- **Dynamic Reality**. Reality is seen as dynamic and ever-changing because beliefs and societal norms continuously evolve.
- **Active Learning:** Constructivism suggests that knowledge is actively constructed by the learner rather than passively absorbed. Learning is seen as a process of adjusting mental models to better fit the world as understood through experience.
- **Relativism:** There is a rejection of the notion of an absolute truth. Instead, multiple truths coexist, depending on perspectives and contexts. Relativism is discussed in depth in Chapter 8.

Constructivism in Action: Consider the role of societal beliefs in shaping responses to climate change. Individuals and communities construct their realities based on media influences, educational backgrounds, and cultural values. As such, one's belief in climate science and the severity of climate change can significantly influence their perception of its reality and urgency. Those who believe strongly in scientific consensus may see climate change as an immediate, tangible crisis, while skeptics, shaped by different beliefs, may perceive the threat as exaggerated or even fabricated.

Overall, the constructivist perspective that beliefs shape reality emphasizes the significant role of cognition, culture, and interaction in forming what individuals accept as reality. It encourages a reflective examination of how perspectives are cultivated and the awareness that multiple realities may coexist within different societal and cultural frameworks.

The Colors of Belief (A Story)

In a small village nestled between the tall mountains and a winding river—much like the setting for the Inn at Mountain Quest situated in the Allegheny Mountains of West Virginia—there lived a young girl named Mira. Mira had a peculiar gift: she could see colors that no one else could. These were not the ordinary colors of the rainbow, although those were often woven between the earth and skies of the valley. These were hues that seemed to come alive, swirling and merging in ways that spoke stories she alone could hear.

The villagers often found Mira's tales of these vibrant, dancing colors amusing, yet puzzling. To them, the world was composed of primary colors—greens from the grass, blues from the sky, and browns from the earth, all beautiful in their own right. But to Mira, the world was alive with glowing whispers entangling themselves with the living energies all around us.

Her mother, a weaver, often listened with a mixture of wonder and confusion as Mira described the colors she saw in the fabrics her mother crafted. "This one is singing with the color of warmth and laughter," Mira would say, holding up a freshly woven scarf. Her mother smiled, indulging the imagination of youth.

Yet, Mira's gift wasn't always celebrated. Some villagers whispered about Mira being too fanciful, her lengthy stories painting her as someone living too much in her dreams. The village elder, known for his stern countenance, often reminded everyone, "Reality doesn't bend to fantasy, my child," casting a stern glance at Mira when she shared one tale too many.

One day, a traveler from a far-off city arrived, bringing with him stories of places where people spoke a hundred languages and the sky glowed with every sunset, with hues Mira had only dreamt of. He was a painter named Luca, searching for new inspirations. When he met Mira, he was intrigued by her vivid descriptions of unseen colors.

"What do you see when you look at these mountains, Mira?" Luca asked her one evening as they stood on the top of the hill overlooking the valley, watching the distant movement of life in the little town.

Mira's eyes widened. "They hum with deep violet, like silence before a storm, mixed with bursts of golden joy."

Luca nodded thoughtfully. "I want to paint that," he said. He began to paint, instead of the static views on the horizon the villagers were used to, he intertwined Mira's vibrant descriptions into sweeping brush strokes. The mountains danced on the canvas, reflecting the world through Mira's eyes.

The villagers, curious about the traveler's strange methods, gathered around to watch. As Luca painted, he shared his philosophy: "Beliefs shape our reality. Mira sees more than colors; she sees the stories the world wishes to tell. And stories are powerful things."

When Luca's painting was completed, the villagers were stunned. The artwork didn't just depict a landscape; it communicated the unseen beauty Mira always spoke of. The canvas was alive with emotions, the colors reaching out to pull on heartstrings. Slowly, eyes that were once skeptical now opened to new perspectives.

Mira became a beloved storyteller, her talents celebrated and embraced. The villagers learned to listen to the colors anew, each shade deepening their understanding of the world around them.

As for Luca, he journeyed onward, carrying with him the inspiration of the colors he couldn't see yet knew existed. For in Mira, he found a truth: belief isn't merely about seeing what others cannot; it's about understanding that the reality we construct holds the power to transform not just visions but lives.

And so, just as the Inn at Mountain Quest has become, the small village by the mountains became a place renowned for its vibrancy—a testament to the idea that when beliefs meet imagination, they paint an inviting reality much richer and more colorful than ever imagined.

Human "Will"

In the superhero movie Green Lantern (Campbell, 2011) based on the DC Comics character of the same name, Hal Jordan, a test pilot, is chosen to become a member of the intergalactic police force called the Green Lantern Corps. The power of the Corps comes from rings that harness the power of will, which is represented by the color green in the emotional spectrum that the movie explores. The ring wearer's willpower turns thought into reality, and thus the stronger one's will the more power they have.

The themes of willpower and courage resonate as timeless lessons about personal growth and the human spirit. First, *overcoming fear*. The film underlines that acknowledging and confronting fear is crucial to understanding and strengthening one's will. Fear can often paralyze or hinder progress, but by addressing it head-on, one can grow more resilient. Second, the importance of *confidence and belief*. Hal's journey shows the importance of self-belief. He makes mistakes and doubts himself, but eventually realizes that believing in his own capabilities is key to fulfilling his potential. Third, in *responsibility and determination*. With great power comes great responsibility, and Hal learns to channel his will to protect what he cares about. Determination can drive actions that align with one's values and responsibilities, even in the face of daunting challenges. Fourth, in *teamwork and trust*. Will is also portrayed as something that can be supported and strengthened by others. The Green Lantern Corps exemplifies how collective will, trust, and cooperation can overcome threats that one alone might not be able to handle.

The concept of "will" plays a fundamental role in shaping the human experience. At its core, will refers to the capacity to make conscious choices, exert control over one's actions, and pursue goals with intention and determination. It is deeply intertwined with notions of desire, motivation, and autonomy. Whether viewed through philosophical, psychological, or neuroscientific lenses, the will is recognized as a driving force behind both individual behavior and broader human endeavors.

The *force* of human will plays an irrefutable role in shaping our lives and defining our paths. *Will is crucial for personal empowerment and autonomy*. It allows individuals to make choices based on their values and beliefs, rather than being passively shaped by external circumstances or internal impulses. Through the exercise of will, people assert their independence and agency, enabling them to shape their life paths intentionally.

Will is the backbone of setting and achieving goals. It provides the necessary drive to initiate actions, sustain effort, and overcome obstacles in pursuit of objectives. Whether it's completing a project, forming a new habit, or pursuing long-term aspirations, the will is what transforms intentions into tangible outcomes.

The concept of will is *central to moral and ethical reasoning*. Being able to choose actions based on ethical considerations necessitates a well-developed will. This dimension allows individuals to align their behaviors with moral principles, contributing to a just and ethical society.

Willpower is critical for self-control and managing impulses. It enables individuals to resist short-term temptations in favor of long-term benefits. This aspect of will is essential for maintaining discipline, forming healthy habits, and avoiding detrimental behaviors.

The human will is *a key factor in resilience*—the capacity to bounce back from challenges and setbacks. A strong will helps individuals to persevere through adversity, adapt to new circumstances, and remain committed to their goals despite difficulties

In everyday life, the will is the silent architect behind countless decisions and actions. It governs how we prioritize tasks, how we respond to challenges, and how we navigate the complex web of personal and professional relationships. Through the exercise of will, individuals craft their identities and contribute to collective societal progress.

The Intertwining of Belief and Will

In the context of philosophy and psychology, "will" is often defined as the *cognitive faculty* by which individuals make choices and decisions, *exert control over their actions*, and *direct intentional behavior*. (See the discussion of "intention" in Chapter 7.) As introduced above, human will is closely tied to notions of agency, autonomy, and motivation, reflecting a person's ability to act according to their desires, goals, or reasons.

Will is central to discussions of free will, moral responsibility, and self-determination, as it underlies one's capacity to initiate and regulate actions in pursuit of desired outcomes. This concept emphasizes the active and intentional aspect of human cognition, contrasting with more passive processes like belief formation or perception, which are often seen as responses to external stimuli or evidence. In exploring the interplay between belief and will, understanding will as a driving force behind deliberate actions helps illuminate why beliefs aren't typically subject to direct voluntary control in the same manner.

Bernard Williams (1973), a prominent philosopher, delved deeply into the interplay between belief and will in his work, particularly in the context of whether it is possible to *willingly decide to believe something*. Williams challenged the notion that beliefs are susceptible to direct voluntary control in the same way that actions are. He argued that belief is inherently truth-directed; it aims at representing the world as it actually is. This intrinsic connection to truth makes deciding to believe something merely by an act of will problematic, because belief cannot simply be willed into existence if it lacks evidence or conviction. (See the discussion of epistemological relativism—truth as knowledge, situation-dependent, context-sensitive, and temporally and individually relevant—in Chapter 8.)

Williams posits that beliefs form as a response to how we perceive and interpret evidence rather than through sheer volition. There is much evidence supporting this. While you cannot directly choose beliefs, you can influence your exposure to evidence and arguments, opening yourself to conviction which might eventually lead to belief alteration.

As can be seen, belief is not as malleable as desire or intention, which can be more directly influenced by will. However, in some instances, individuals might persuade themselves to "believe" in something due to pragmatic reasons—such as the comfort or hope a belief might bring. Williams would assert that this is not true belief but rather a form of self-deception or wishful thinking.

The intertwining of belief and will resonates with the theme of not seeing what is directly before us by emphasizing how cognitive biases and motivated reasoning can shape our perception and understanding of reality. When individuals decide to believe something because it aligns with their desires, or dismiss evidence contrary to their preferred beliefs, they demonstrate a form of epistemic blindness. This selective attention and resistance to counter-evidence can lead individuals to overlook or misinterpret what is immediately apparent, essentially shaping an alternate reality based on personal biases rather than objective truth.

Let's specifically explore the intertwining of belief, will, and reality creation in terms of constructed reality, cognitive dissonance, collective influence, and narrative construction.

Constructed Reality: Just as constructivist theories propose (see above), belief and will converge to construct a version of reality tailored by personal biases, motivations, and cultural contexts. When an individual wills themselves to adopt certain beliefs—whether through indoctrination, societal influence, or self-deception—they contribute to creating a subjective lens through which reality is perceived and interpreted.

Cognitive Dissonance: When reality contradicts a willed belief, cognitive dissonance may arise. To resolve this discomfort, individuals might alter their perception or selectively attend to evidence that supports their belief, further intertwining belief and will. (See Chapter 11.)

Collective Influence: On a societal level, the confluence of belief and will can lead to shared realities, such as cultural or institutional beliefs that persist despite contradicting evidence. For instance, societal norms or ideologies might prevail even when they're not grounded in empirical truth, highlighting the powerful role of collective will in shaping perceived realities.

Narrative Construction: Individuals craft narratives that help them navigate and make sense of the world. These narratives, often shaped by a combination of belief and will, dictate what people see, cherish, or discard in their perceived reality. (See Chapter 13 under "Psychology".)

Williams' research provides a critical foundation for examining how belief intersects with will, particularly in how it influences perception and the construction of reality. Understanding this interplay is crucial for recognizing how individuals can be blind to evident truths and how collective belief systems can shape societal realities, often reinforcing shared narratives that may not align with objective or empirical evidence.

From Passivity to Agency

To realize, perhaps suddenly, that we participate in the creation of our reality can evoke intense joy—a palpable burst of freedom and possibility. No longer hemmed in by the need to simply "fit in," one discovers the playful aspect of existence: "What if I could choose to see—and act—differently?" Carol Dweck's research (2006) identifies two basic "mindsets" that people use to interpret their abilities and potential: the "fixed" mindset (believing traits like intelligence or talent are static) and the "growth" mindset (believing these traits can be developed through effort and learning). Her findings reveal that adopting a growth mindset transforms how people interpret challenges and failure—reality becomes fluid and full of possibilities for development, fueling motivation, resilience, and a sense of agency. For example, a student who, upon discovering that "intelligence" is not fixed but can be grown, reimagines themselves as an infinite learner. The burden of inadequacy transforms into the joy of growth, experimentation, and personal design.

This dawning realization brings a spirit of experimentation. If meaning is made, not given, then life becomes an open field for interpretation, artistry, and self-renewal. Constructivist educators celebrate this playful spirit as central to deep learning. John Dewey's (1933) classic educational philosophy centers on the idea that thinking is not passive reception but an *active process* of interpretation and growth. Reflective thinking, as Dewey describes it, involves questioning assumptions, testing possibilities, and learning from experience by reorganizing one's understanding. Education and life are both about continual reconstruction of meaning, emphasizing the creative, participatory nature of knowing and being.

The freedom to reframe can turn defeat into opportunity, isolation into autonomy, and confusion into curiosity. Gross and John (2003) examine how people's habits of "cognitive reappraisal" (re-framing the meaning of situations) and "suppression" (inhibiting emotional responses) affect their well-being and relationships. They find that cognitive reappraisal—actively choosing how to interpret an event—increases happiness and life satisfaction, confirming that the stories we tell ourselves about experience can change our felt reality and our agency within it.

This doesn't mean all realities are equally easy or safe to change, but within limits, new stances are possible—and this is exhilarating. For example, after years in a rigid workplace, an employee transforms their relationship to daily routines by choosing to focus on small opportunities for kindness and creativity, discovering joy in what once felt monotonous.

Accepting one's role in reality-creation does not mean ignoring the world's constraints or denying community, but weaving what is given with what can be made. Joy arises not from limitless control, but from agency—the capacity to participate, to reframe, and to re-make oneself and one's world in partnership with others and with circumstance. This recognition also deepens empathy: if I create meaning, so do others. Our realities both overlap and diverge, and dialogue becomes a creative encounter.

For example, Hans-Georg Gadamer (1975/2004) explores how understanding is formed—not in isolation, but through the interplay ("fusion of horizons") between our own perspectives and those of others, past and present. He insists that truth emerges dialogically, as we bring our questions and experiences into conversation with different traditions, texts, and viewpoints, illuminating how reality is creatively shaped and reshaped as we engage openly and responsively with other perspectives.

To awaken to the truth that we actively shape our realities—alone and with others—is to enter a field of possibility charged with joy, agency, and meaning. This is the freedom described by philosophers, the resilience celebrated by psychologists, and the creativity cultivated by artists and teachers. Life becomes a canvas, and we its ever-learning painters.

Creating Our Reality (An Exercise)

Creating one's reality through the interplay of "will" and "beliefs" involves a thoughtful and conscious approach to shaping perceptions and interactions with the world. Here's a step-by-step process that outlines this integration, inclusive of a scenario example. Related to our core focus, the goal is to overcome perceptual barriers that prevent us from recognizing possibilities, truths, or solutions readily available, right in front of us.

Scenario Context: Luca, a manager at a software company, who wants to improve their team's productivity by recognizing and acting on issues that are right in front of them but have been ignored.

Step 1: Self-Awareness. *Reflect on Current Beliefs:* Begin with introspection to identify existing beliefs. Consider how these beliefs shape your perception of reality and whether they serve your well-being or goals. *Acknowledge Emotional Responses:* Notice how your emotions are tied to your beliefs. Emotions can be strong indicators of deeply held convictions that influence your reality. *Acknowledge Blind Spots:* Recognize that everyone has perceptual blind spots influenced by entrenched beliefs. Use self-awareness to identify areas where you might be ignoring or misinterpreting obvious information. *Reflect on Cognitive Biases:*

Understand common cognitive biases (such as confirmation bias) that may prevent you from seeing what's right in front of you and commit to questioning these biases.

Scenario Example: (*Identify Blind Sports*) Luca reflects on frequent complaints about communication issues within the team but realizes they haven't acknowledged these as a significant barrier because they believe productivity struggles are due to team skills, not communication. (*Understand Cognitive Biases*) After learning about confirmation bias, Luca admits they've been focusing only on feedback that aligns with their belief that the issue lies in technical training, rather than communication.

Step 2: Identify Desired Reality. *Define Your Goals:* Clearly articulate what you want to achieve or how you wish to experience your life. *Visualize Outcomes:* Imagine in detail the reality you wish to create. Visualization helps to establish a concrete sense of direction. *Set Intentions to See Clearly:* Clearly articulate an intention to remain open to new perspectives, seeking to rid yourself of assumptions that cloud clear vision. *Practice Curiosity:* Cultivate a mindset of curiosity, which involves actively seeking out new information or viewpoints contrary to your current beliefs.

Scenario Example: (Set Intentions to See Clearly) Luca sets a clear intention to actively notice and address the underlying communication issues rather than attributing everything to skill deficits. (*Practice Curiosity*) Luca decides to approach team interactions with curiosity, asking open-ended questions to uncover insights into team dynamics and communication patterns.

Step 3: Evaluate and Challenge Beliefs. *Question Limiting Beliefs:* Identify beliefs that may be hindering your desired reality and critically assess their validity. *Seek Evidence:* Look for evidence that contradicts limiting beliefs and consider alternative perspectives that support new, empowering beliefs. *Engage in Critical Thinking:* Regularly challenge your beliefs with critical thinking. Ask probing questions and seek evidence that might disprove your current assumptions, allowing new insights to surface. *Seek Disconfirming Evidence:* Make a conscious effort to seek out information that contradicts your existing beliefs or reinforcing evidence. This helps in broadening perception and uncovering overlooked truths.

Scenario Example: (*Engage in Critical Thinking*) Luca starts questioning their belief that technical skills are the primary productivity enhancer and seeks evidence that communication could play a pivotal role. (*Seek Disconfirming Evidence*) Luca reviews past feedback and notes specific instances where poor communication resulted in project delays, indicating an overlooked reality.

Step 4: Exercise Will to Influence Belief. *Cultivate Openness:* Use willpower to consciously expose yourself to diverse perspectives and new information, which may help reshape beliefs. Humility is a human choice that helps being open to listening to others' ideas and viewing the world from different perspectives. *Practice Affirmations:* Regularly affirm beliefs aligned with your desired reality. Affirmations can slowly influence your subconscious mind, promoting belief change. *Be Mindfully Present:* Use your will to practice mindfulness, staying present in the moment to fully engage with the environment and the people around you, which can reveal overlooked details. *Focus on Sensory Awareness:* Heighten your sensory awareness through exercises that encourage deeper engagement with your immediate surroundings, enhancing your ability to notice subtleties you might have missed.

Scenario Example: (*Be Mindfully Present*) In team meetings, Luca focuses on being present, tuning into interpersonal exchanges and subtleties in communication that were previously ignored. (*Focus on Sensory Awareness*) To improve perceptiveness, Luca practices active listening, paying close attention to tone, body language, and engagement levels during discussions.

Step 5: Align Actions with Beliefs. *Take Consistent Action.* Ensure your actions consistently reflect your new beliefs and the reality you wish to create. *Practice Mindfulness.* Stay present and conscious of daily actions and decisions to ensure alignment with your consciously chosen beliefs. *Implement Adaptive Strategies:* Actively incorporate feedback and changes into your behavior based on newly perceived information, reinforcing the ability to adapt and see the obvious. *Cultivate a Growth Mindset:* Adopt a growth mindset that encourages viewing challenges and new information as opportunities for personal development and clearer perception.

Scenario Example: (*Implement Adaptive Strategies*) Luca gathers the team for a workshop to enhance communication skills and introduces new tools and practices to facilitate better information sharing. (*Cultivate a Growth Mindset*) Luca fosters an environment where team members perceive feedback and communication improvements as part of personal and professional growth.

Step 6: Engage with Supportive Environments. *Surround Yourself with Positivity:* Seek out environments and communities that reinforce your chosen beliefs. *Communicate Your Vision:* Share your goals and beliefs with supportive individuals who can provide encouragement and accountability. *Seek Diverse Interactions:* Surround yourself with a diverse group of people who offer different perspectives, increasing your exposure to varied viewpoints and challenging entrenched beliefs. *Discuss and Reflect:* Engage in discussions that encourage reflection and re-evaluation of previously held beliefs, fostering an environment conducive to seeing what's been missed.

Scenario Example: (*Seek Diverse Interactions*) Luca invites feedback from various team members about communication preferences and challenges to gather a comprehensive view. (*Discuss and Reflect*) Regular team debriefs are scheduled, allowing the team to express issues and successes openly, encouraging reflective thinking and problem-solving.

Step 7: Adapt and Evolve. *Stay Flexible:* Be willing to adapt beliefs and actions as new information and experiences arise. *Reflect Regularly:* Periodically revisit and reassess your beliefs and goals to ensure they continue to align with your evolving reality. *Regular Reassessments:* Periodically reassess your beliefs and perceptions to ensure they reflect current realities and insights, adjusting your approach as necessary. *Celebrate Discoveries:* Acknowledge and celebrate the moments when you do notice what's been overlooked, reinforcing the value of sustained effort to see clearly.

Scenario Example: (*Regular Reassessment*) Luca sets up monthly check-ins to reassess team dynamics and communication effectiveness, ensuring any new blind spots are identified and addressed. (*Celebrate Discoveries*) Luca openly acknowledges improvements in communication and how it has positively affected productivity, celebrating these wins with the team.

Step 8: Acknowledge and Integrate Learning. *Learn from Setbacks:* Accept that challenges and setbacks are part of growth. Use them as opportunities to learn and strengthen your belief system. *Will Yourself to Become an Infinite Learner:* Recognize that as a complex adaptive system, learning is living and living is learning, treating learning as you would a religion, embracing all human capacities and opportunities to learn. *Celebrate Progress:* Recognize and celebrate the shifts you make in creating your desired reality, reinforcing the positive impact of aligned belief and will.

Scenario Example. (*Infinite Learning*) Luca accepts the Infinite Learning challenge: to keep the "individual" in *continuous* and *lifelong learning*, to retain the "social" in *social learning*, to embrace the value (often tacit and intuitive) of *unconscious learning*, to explore the potential in *quantum cognition* (the energetic flow of information), to honor and fully integrate into his everyday actions the *learning journey of our soul* (the spiritual dimension), and to take full advantage of the incredible potential offered through *human mind/brain capacity and capability* (Bennet, 2025).

By incorporating these steps, individuals can leverage the power of will and belief to break down perceptual barriers, enabling them to see what's directly in front of them with greater clarity and understanding. Luca became more aware of communication issues previously overlooked, and also effectively addressed them, leading to a more cohesive and productive team environment. This iterative approach helped overcome perceptual blind spots by actively engaging belief, will, and action in a continuous feedback loop. This approach is about fostering a dynamic perception that adapts and expands rather than remaining stagnant and limited.

Collective Reality Creation

Our individual ability to shape reality reaches its fullest potential when joined with the creative capacities of others. The journey from "I" to "We" in reality creation illuminates how communal perspectives enhance not only personal awareness but also collective understanding and growth. While we've touched on the intricate ways individuals craft their personal realities, we now pivot to the shared space where "We" create a world more vividly and expansively than apart.

In the hustle of life, and especially in the midst of the chaotic and unpredictable todays, it's easy to miss what lies right in front of us, partly because our selective attention filters individual experiences, as discussed extensively in previous chapters. Yet the individual next to you may be seeing an entirely different reality, with their own "blind spots" that are visible to you. Each of these narrow lenses can widen through dialogue, shared ventures, and a collective commitment to re-envisioning the world. When we come together to form a collective reality, the limitations of individual perception give way to a richer tapestry, woven from diverse strands of human insight.

At the core of collective reality creation is intersubjectivity—the mutual sharing and shaping of perceptions and meanings. As individuals, we only see fragments of reality, often clouded by personal biases and assumptions. By joining in conversation with others, we open ourselves to new dimensions of understanding, collective insight, and a reimagining of what's possible. As Berger and Luckmann (1966) have shown, reality is constructed through ongoing human interaction. This process is vital in understanding why we can't see everything that's in front of us until it's collectively revealed.

For example, the workplace, often cited as a focal point for stress and conformity, can transform in the image of the community's combined visions and efforts. If one person reimagines their role, they begin a ripple effect. However, when this mindset infects a team or an organization, the ripple becomes a wave. Group dynamics morph, reshaping routines and cultures into ones that emphasize creativity, collaboration, and mutual empowerment. This transformation is not seen until each individual acts and engages, making the invisible challenges of work turn into visible opportunities for change, perhaps embedded into the culture, stretching that learning toward the future.

Creating our reality is not about dominating what is present, but about weaving the already-given with the yet-to-be-imagined, together. Recognizing our shared role in building the world also deepens collective responsibility; we become aware that our co-created reality must be tended to with care, understanding, and respect for the diversity of human experience. When collective agency is exercised with an awareness of difference, empathy becomes the bridge across which reality is jointly constructed.

Hans-Georg Gadamer (2004) speaks to the importance of "fusion of horizons"—the idea that understanding and truth are born out of the interplay between our perspectives and those of others. This concept encapsulates the heart of moving from "I" to "We" in reality creation. In conversations marked by openness and creative engagement, our individual horizons merge, allowing us to see what was previously hidden, and revealing deeper truths about ourselves and our communities.

As we shape reality together, we participate in a grander narrative of human evolution, becoming stewards of both collective joy and collective growth. The challenges we face individually or communally become opportunities for shared reinvention. Within this participatory framework, we realize that our combined creativity fosters new possibilities for living, working, and being together.

Ultimately, transitioning from "I" to "We" in creating our reality reveals the fundamental truth that life is not a solitary journey. As constructs of reality are shared, they become more vibrant, complex, and enduring. Individually, we might miss the fullness of what lies in front of us. Together, however, our shared vision can illuminate paths unseen, breathing life into the spaces in which we coexist. Through this collective creativity, we achieve a more profound comprehension of our place in the world and our potential to contribute to its ongoing creation. By embracing "We," we transcend our individual limitations, inviting joy and discovery as the hallmarks of our shared human experience

Final Thought

When it's all said and done, the reality in front of us, what we "see" right in front of us, is what we *want* to see, what we *choose* to see, individually and collectively. Yet on reflection, and with our new understanding of how we create our reality, the *whole world is right in front of us*—a field full of possibilities.

Choose Whysly.

Appendix A: Socratic Questioning

Socratic questioning is a form of disciplined questioning that Socrates, the ancient Greek philosopher, used to stimulate critical thinking and illuminate ideas. This method involves asking guiding questions to encourage deeper reflection and self-discovery, rather than just delivering answers or information. The overarching goal of Socratic questioning is to promote critical thinking, clarify concepts, and uncover underlying beliefs and assumptions.

Key features of Socratic questioning include encouraging dialogue, focusing on understanding, promoting reflection, challenging assumptions, and seeking clarity.

Encourages Dialogue: Socratic questioning fosters a dialogue instead of a debate. It invites participants to explore their thoughts and feelings openly.

Focuses on Understanding: The emphasis is on understanding the "why" and "how" of beliefs and ideas, rather than just the "what." This helps to uncover deeper meanings.

Promotes Reflection: The emphasis is on understanding the "why" and "how" of beliefs and ideas, rather than just the "what." This helps to uncover deeper meanings.

Challenges Assumptions: The emphasis is on understanding the "why" and "how" of beliefs and ideas, rather than just the "what." This helps to uncover deeper meanings.

Seeks Clarity: The method aims to clarify thoughts and ensure that concepts and terms are understood in the same way by all parties involved in the discussion.

There are five types of questions in Socratic Questioning. First, *clarification questions* such as "What exactly do you mean by that?" or "Can you give me an example?" Second, *probing assumptions* such as "What are you assuming here?" or "How did you come to that conclusion?" Third, *exploring reasons and evidence* such as "What evidence supports your claim?" "Are there alternative viewpoints we should consider?". Fourth, *considering implications and consequences* such as "What happens if we accept this idea?" or "What are the potential consequences of this action?" And fifth, *questioning the question* such as "Why is this question important?" or "What are we trying to achieve by asking this?"

Socratic questioning is widely used in various fields, including education, therapy, and critical thinking exercises. In education, teachers might use it to engage students in deeper discussions about a subject, helping them evaluate their knowledge more critically. In therapy, it can help individuals explore their thoughts and beliefs, achieve greater insight, and facilitate personal growth.

By employing Socratic questioning, individuals can develop a habit of critical thinking and refine their ability to articulate their thoughts and beliefs. This technique promotes intellectual rigor and fosters an open-minded approach to understanding complex issues, making it a valuable tool for education, discussion, and personal development.

A Socratic Questioning Example

Let's apply the Socratic questioning methodology to the core question of this book: "Why can't I see what is right in front of me?"

1. Clarification Questions

"What exactly do you mean by that?" In this context, it is valuable to clarify what "seeing" refers to. Are you referring to a literal visual perception, or does it relate to understanding a situation or an idea? [In this case, the verb "seeing" was defined in the beginning of Chapter 1 as both literal and metaphorical, representing sight as well as noticing, perceiving or comprehending something beyond visual observation.] Additionally, what specifically is it that you *feel* is right in front of you? [This is dependent on the individual.]

"Can you give me an example?" Defining what is meant by "right in front of me" can help narrow it down and avoid ambiguity. [Something that is happening in my reality that I cannot perceive or even understand.] This clarification could involve both physical objects and more abstract concepts such as opportunities, emotions, or relationships. It could also relate to the larger fragmentation of the current environment.

In a work-related environment, an example could be a situation where you're in a meeting and fail to notice a colleague's non-verbal cues indicating discomfort. Another example might be missing obvious solutions to a problem at work because you are too focused on a single approach. These instances illustrate how one can overlook what's directly in front of them.

2. Probing Assumptions

"What are you assuming here?" The assumption might be that there is something clear and obvious in front of you that you should be able to see. Are you presuming that you are entirely objective in your perspective? [As is reinforced throughout the book, this is not the case.] It may also imply a belief that awareness and perception are purely physical or cognitive processes when they can also be influenced by emotional state, biases, or distractions. Identifying these assumptions can help uncover barriers to your perception and understanding.

"How did you come to that conclusion?" Recognizing that there is something in front of you that you do not "see" could be drawn from past experiences wherein you recognized similar patterns of overlooking information or failing to notice interpersonal dynamics. Additionally, current world events, self-reflection or feedback from others may have led you to question your situational awareness.

3. Exploring Reasons and Evidence

"What evidence supports your claim?" To address the inability to see what's in front of you, consider the reasons behind this claim. Is there specific evidence, like, as mentioned above, past experiences where you overlooked something obvious, that supports this assertion? You might also reflect on instances where your judgment or perception has been clouded by emotions or preconceived notions. Exploring alternative viewpoints—such as perspectives from friends or mentors—who might see things differently could provide insight and help validate or challenge your claims.

"Are there alternative viewpoints we should consider?" Yes, perhaps it's useful to consider the viewpoint that maybe what's in front of you is not as clear as it seems and that others perceive it differently. Understanding that different contexts and personal biases can influence perception might provide a more nuanced perspective on your situation. What are your biases? And, when you look around your environment, and see how others are responding to this environment, what are you missing? What learning needs to occur so that you can better understand the reality in which you are living? For example, in a bipartisan government, what are other points of view that you have not considered? How are the choices being made today affecting you directly?

4. Considering Implications and Consequences

"What happens if we accept this idea?" If the idea that you can't see what's in front of you is accepted, it could imply that there are limitations to your perception or awareness that need addressing. This may lead to a deeper understanding of your thought processes and emotional state, potentially sparking a quest for greater self-awareness. Understanding these limitations could result in making more informed decisions in the future or motivate a change in behavior, such as seeking external perspectives to broaden your outlook.

"What are the potential consequences of this action?" The consequences could range from improving your ability to recognize opportunities and challenges to potentially increasing feelings of frustration or anxiety if you realize a pattern of overlooked awareness. However—if you *choose whysly*—this latter realization can also lead to personal growth and improved relationships as you become more attuned to your environment.

6. Questioning the Question

"Why is this question important?" This question is significant because it touches on the essence of self-awareness, perception, and understanding. It challenges you to reflect on your cognitive processes and emotional state and invites exploration into personal biases or distractions that may be obstructing your view. By investigating this question, you can unlock insights into aspects of your life that may need attention or improvement, paving the way for personal growth and improved clarity and choosing whysly in decision-making.

"What are we trying to achieve by asking this?" By asking this, we're aiming to achieve a deeper understanding of personal cognitive or emotional blind spots, improve situational awareness, and foster meaningful dialogues that promote clarity and insight in various aspects of life. And ultimately, we're trying to increase individual awareness of the capacity to "see", understand, and reflect on the changes underway in the larger world and our ability to influence and direct those changes toward a positive return to humanity.

As can be seen, the Socratic questioning methodology can be an effective tool for reflection and facilitating a deeper understanding of the underlying reasons for our inability to "see" what is right in front of us.

References

Althusser, L. (1971). Ideology and ideological state apparatuses. In L. Althusser, *Lenin and philosophy and other essays* (B. Brewster, Trans.), pp. 127–188. Monthly Review Press.

Altmann, G. T. M. (1997). *The ascent of Babel: An exploration of language, mind, and understanding*. Oxford University Press.

Anderson, J. R. (2010). *Cognitive psychology and its implications* (7th ed.). Worth Publishers

Aquinas, T. (1911-1925). *Summa theologica* (Fathers of the English Dominican Province, Trans.). Benziger Brothers. (Original work written 1265-1274)

Ariely, D. (2008). *Predictably Irrational: The Hidden Forces That Shape Our Decisions*. HarperCollins.

Ariely, D. (2012). *The (honest) truth about dishonesty*. Harper.

Aristotle. (2009). *Nicomachean ethics* (W.D. Ross, Trans.). Oxford University Press. (Original work published c. 350 B.C.E.)

Asch, S. E. (1955). "Opinions and social pressure." *Scientific American* , 193(5), 31-35. https://doi.org/10.1038/scientificamerican1155-31

Astin, J. A., Berman, B. M., & Wallace, J. (2003). The efficacy of mind-body therapies in the treatment of migraine: A systematic review. *The Clinical Journal of Pain*, *19*(1), 1-10. https://doi.org/10.1097/01.ajp.0000083023.68189.26

Atran, S., & Medin, D. (2008). The cultural mind: Environmental decision-making and cultural models . In *The Handbook of Cultural Psychology* (pp. 310-329). Guilford Press.

Atran, S., & Medin, D. (2008). The cultural mind: Environmental decision-making and cultural models . In *The Handbook of Cultural Psychology* (pp. 310-329). Guilford Press

Baars, B. J. (1988). *A cognitive theory of consciousness*. Cambridge University Press.

Baghramian, M. (2019). *Relativism*. Stanford Encyclopedia of Philosophy.

Banaji, M. R., & Greenwald, A. G. (2013). *Blindspot: The hidden biases of good people*. Delacorte Press.

Barkley, R. A. (2015). *Attention-deficit hyperactivity disorder: A handbook for diagnosis and treatment* (4th ed.). Guilford Press.

Bartlett, F. C. (1932). *Remembering: A study in experimental and social psychology*. Cambridge University Press.

Baumeister, R. F., & Exline, J. J. (2000). *Identity, self, and social comparison*. In S. T. Fiske & D. T. Gilbert (Eds.), *Handbook of Social Psychology* (4th ed.), pp. 680-707.

Baumeister, R. F., Bratslavsky, E., Finkenauer, C., & Vohs, K. D. (2001). Bad is stronger than good. *Review of General Psychology*, *5*(4), 323-370.

Beall, J. C., & Restall, G. (2006). *Logical pluralism*. Oxford University Press.

Beauvais, C., & Whelchel, H. (2014). Cognitive dissonance and religious change: A study of the relation between cognitive dissonance and religious conversion. *Sociology of Religion*, *75*(3), 392-412.

Beck, J. S. (2011). *Cognitive Behavior Therapy: Basics and Beyond* (2nd ed.). Guilford Press.

Bell, C. (1992). *Ritual theory, ritual practice*. Oxford University Press.

Benedict, R. (1934). *Patterns of culture*. Houghton Mifflin.

Bennet, A. (2018). *Possibilities that are YOU! : Vol. 4 conscious compassion*. MQIPress.

Bennet, A. (2025). *Contiguity: Entangled living and learning*. MQIPress.

Bennet, A. (2025a). *Infinite learning: Your life, your choice*. MQIPress.

Bennet, A. (2025b). *fRAGmentation : The power of e pluribus unum*. MQIPress.

Bennet, A. (2025c) *The taint of hubris: Rediscovering humility and bridging authentic connection*. MQIPress.

Bennet, A. (2025d) *Contiguity: Entangled living and learning*. MQIPress.

Bennet, A. (2025e) *Choosing whysly: Why we can't see what is right in front of us*. MQIPress.

Bennet, A., & Scott, C. L. (2022). *With passion, we live and love: Research, prose, verse and music*. MQIPress.

Bennet, A., & Shelley, A. with Dhewa, C. (2024). *Innovative creativity: Creating with innovation in mind*. MQIPress.

Bennet, A., & Turner, R. (2023). *Reblooming the knowledge movement: The democratization of organizations*. MQIPress.

Bennet, A., & Turner, R. (2024a). *Whole thought: The rise of human intelligence. MQIPress*.

Bennet, A., & Turner, R. (2024b). *Knowledge capacities I: Igniting whole thought*. MQIPress.

Bennet, A., & Turner, R. (2025). *C&C thinking (critical and creative): Becoming whole*. MQIPress.

Bennet, D., Bennet, A., & Turner, R. (2018). *Expanding the self: The intelligent complex adaptive learning system*. MQIPress.

Bennet, D., Bennet, A., & Turner, R. (2022). *Unleashing the human mind: A consilience approach to managing self.* MQIPress.

Berger, P. L., & Luckmann, T. (1966). *The social construction of reality: A treatise in the sociology of knowledge*. Penguin Books.

Bloor, D. (1976/1991). *Knowledge and social imagery* (2nd ed.). University of Chicago Press.

Blumer, H. (1969). *Symbolic interactionism: Perspective and method*. University of California Press.

Boas, F. (1940). *Race, language, and culture*. Macmillan.

Boghossian, P. (2006). *Fear of knowledge: Against relativism and constructivism*. Oxford University Press.

Bohm, D. (1996). *On dialogue.* Routledge

Boroditsky, L. (2001). Does language shape thought?: Mandarin and English speakers' conception of time. *Cognitive Psychology*, *43*(1), 1-22.

Broadbent, D. E. (1958). *Perception and communication*. Pergamon Press.

Brown, B. (2018). *Braving the wilderness: The quest for true belonging and the courage to stand alone.* Random House.

Brown, K. W., & Ryan, R. M. (2003). The benefits of being present: Mindfulness and its role in psychological well-being. *Journal of Personality and Social Psychology* , *84*(4), 822-848. https://doi.org/10.1037/0022-3514.84.4.822

Brown, K. W., Ryan, R. M., & Creswell, J. D. (2007). Mindfulness: Theoretical foundations and evidence for its salutary effects. *Psychological Inquiry*, *18*(4), 211-237.

Buckner, R. L., Andrews-Hanna, J. R., & Schacter, D. L. (2008). The brain's default network: Anatomy, function, and relevance to disease. *Annals of the New York Academy of Sciences, 1124* (1): 1–38.

Calvert, G., Spence, C., & Stein, B. E. (2004). *The Handbook of multisensory processes*. MIT Press.

Campbell, M. (Director). (2011). *Green lantern* (film). Warner Bros. Pictures.

Chui, M., Manyika, J., & Miremadi, A. (2016). Where machines could replace humans—and where they can't (in the near term). *McKinsey Quarterly*. Retrieved from https://www.mckinsey.com/featured-insights/future-of-work/where-machines-could-replace-humans-and-where-they-cant-in-the-near-term

Cisek, P., & Kalaska, J. F. (2010). Neural correlates of reaching movements in the dorsal and ventral streams of the primate motor system. *Cerebral Cortex* , *20*(2), 251-262. https://doi.org/10.1093/cercor/bhp097

Clark, A. (2013). Predictive brains, situated agents, and the future of cognitive science. *Behavioral and Brain Sciences, 36*(3), 181-204.

Clark, A., & Chalmers, D. J. (1998). The extended mind. *Analysis*, 58(1), 7-19.

Cohen, L. E. (2001). "Creating a climate for change: The role of social factors in the development of antiracist coalitions." *Social Justice*, *28*(3), 21-43.

Coull, J. T., & Nobre, A. C. (2008). Dissociating temporal and spatial attention: Evidence from event-related brain potentials. *Cognitive Neuroscience*, *20*(5), 94-101. https://doi.org/10.1162/jocn.2007.20107

Crick, F. (1994). *The astonishing hypothesis: The scientific search for the soul*. Scribner.

Csikszentmihalyi, M. (1993). The evolving self: A psychology for the third millennium. HarperCollings Publishing.

Damasio, A. R. (1994). *Descartes' error: Emotion, reason, and the human brain*. G.P. Putnam's Sons.

Davidson, D. (Ed.) (1980). *Essays on actions and events*. Clarendon Press.

Davidson, Donald. (1974). "On the Very Idea of a Conceptual Scheme." *Proceedings and Addresses of the American Philosophical Association*, *47*, 5-20.

Dehaene, S. (2014). *Consciousness and the brain: Deciphering how the brain codes our thoughts*. Viking

Dehaene, S., & Changeux, J. P. (2011). Experimental and theoretical approaches to conscious processing. *Nature Reviews Neuroscience*, *12*(5), 453-468. https://doi.org/10.1038/nrn3021

Desjarlais, R. (1992). *Body and emotion: The aesthetics of illness and healing in the Nepal Himalayas*. University of Pennsylvania Press.

Dewey, J. (1916/1933) *Democracy and education*: An introduction to the philosophy of education. Macmillan.

Dewey, J. (1933). *How we think: A restatement of the relation of reflective thinking to the educative process*. Heath and Company.

Dewey, J. (1939). Theory of valuation. In *A logic: The theory of inquiry*. Henry Holt and Company.

Doidge, N. (2007). *The brain that changes itself: Stories of personal triumph from the frontiers of brain science*. Viking.

Doshi-Velez, F., & Kim, B. (2017). Towards a rigorous science of interpretable machine learning. arXiv preprint arXiv:1702.08608.

Dufrenne, M. (1973). *The phenomenology of aesthetic experience*. Northwestern University Press.

Dunning, D., & Kruger, J. (1999). Unskilled and unaware of it: How difficulties in recognizing one's own incompetence lead to inflated self-assessments. *Journal of Personality and Social Psychology*, *77*(6), 1121-1134.

Dunning, D., & Story, A. (1991). The interpersonal implications of self-deception: Multiple influences on the perception of honesty. *Journal of Personality and Social Psychology*, *61*(5), 927.

Dunning, D., Heath, C., & Suls, J. (2004). Flawed self-assessment: Implications for health, education, and the workplace. *Psychological Science in the Public Interest*, *5*(3), 69-106.

Dweck, C. S. (2006). *Mindset: The new psychology of success*. Random House.

Easton, D., & Hess, R. d. (1962). The child's political world. *Midwest Journal of Political Science* *6*(3), 229-246.

Entman, R. M. (1993). Framing: Toward clarification of a fractured paradigm. *Journal of Communication, 43*(4), 51–58.

Eysenck, M. W., & Keane, M. T. (2015). *Cognitive Psychology: A Student's Handbook* (7th ed.). Psychology Press.

Feldman, H. (2013). The predictive brain as a stochastic organ. *Biology Letters, 9*(6), 20130632.

Festinger, L. (1957). *A theory of cognitive dissonance*. Stanford University Press

Festinger, L., & Maccoby, E. (1964). On understanding the relevance of cognitive dissonance to change in attitude. *Advances in Experimental Social Psychology*, *1*, 75-134.

Festinger, L., Riecken, H. W., & Schachter, S. (1956). *When prophecy fails: A social and psychological study of a modern group that predicted the destruction of the world.* Harper and Row.

Feyerabend, P. (1975). *Against method: Outline of an anarchistic theory of knowledge*. Verso.

Field, H. (2009). *Saving truth from paradox* . Oxford University Press.

Fingarette, H. (1969). *Self-deception*. University of California Press.

Flavell, J.H. (1979). Metacognition and cognitive monitoring: A new area of cognitive-developmental inquiry. *American Psychologist, 34*(1), 906-922.

Frankl, V. E. (2006). *Man's Search for Meaning*. Beacon Press.

Freire, P. (2000). *Pedagogy of the oppressed* (30th anniv. ed.). Continuum.

Freud, S. (1923). *The ego and the id*. Standard Edition, Volume 19. Hogarth Press.

Fried, I., & Wilson, C. L. (2008). Neural correlates of intention in a brain-computer interface. *Nature Neuroscience, 11*(12), 1514-1523. https://doi.org/10.1038/nn.2223

Friston, K. (2005). A theory of cortical responses. *Philosophical Transactions of the Royal Society B: Biological Sciences, 360* (1456), 815-836. https://doi.org/10.1098/rstb.2005.1622

Friston, K. (2010). The free-energy principle: A unified brain theory? *Nature Reviews Neuroscience, 11*(2), 127–138.

Gadamer, H. G. (1975-1960). *Truth and method*. Sheed & Ward.

Galbraith, J. K. (1958). *The affluent society.* Houghton Mifflin.

Gazzaniga, M. S. (2018). *The consciousness instinct: Unraveling the mystery of how the brain makes the mind*. Farrar, Straus and Giroux.

Geertz, C. B. (1973). The interpretation of cultures: Selected essays (Carrol, Ed.). Basic Books. MIT Press.

Geirhos, R., Jacobsen, J., Michaelis, C., et al. (2020). Shortcut learning in deep neural networks. *Nature Machine Intelligence*, 2(11), 665-673.

Gerbner, G., & Gross, L. (1976). Living with television: The violence profile. *Journal of Communication, 26*(2), 172–199.

Gibson, J. J. (1979). *The ecological approach to visual perception*. Houghton Mifflin

Gidron, N., & Bonikowski, B. (2013). *Varieties of populism: Literature review and research agenda*. Working Paper.

Goffman, E. (1974). *Frame analysis: An essay on the organization of experience*. Harper & Row.

Goleman, D. (1995). *Emotional Intelligence.* Bantam Books.

Goodfellow, I. J., Shlens, J., & Szegedy, C. (2015). Explaining and harnessing adversarial examples. arXiv:1412.6572.

Gramsci, A. (1971). *Selections from the prison notebooks*. (Ed. and Trans. by Q. Hoare and G. Nowell-Smith). International Publishers.

Grandin, T., & Barron, S. (2005). *The unwritten rules of social relationships: Decoding social mysteries through the unique perspectives of autism*. Future Horizons.

Greene, J. D., & Paxton, J. M. (2009). Patterns of neural activity associated with honest and dishonest moral decisions. *Proceedings of the National Academy of Sciences*, *106*(30), 12506-12511.

Greenfield, S. (2000). *The private life of the brain: Emotions, consciousness, and the secret of the self.* John Wiley & Sons.

Gregory, R. L. (1970). *The intelligent eye*. Weidenfeld & Nicolson.

Gross, J. J., & John, O. P. (2003). "Individual differences in two emotion regulation processes: Implications for affect, relationships, and well-being." *Journal of Personality and Social Psychology,* 85(2), 348–362.

Guthrie, W. K. C. (1971). *The Sophists* . Cambridge University Press.

Hall, S. (1980). Encoding/decoding. In *Culture, Media, Language: Working Papers in Cultural Studies, 1972-79*, 128-138. Routledge.

Hall, S. (1997). *Representation: Cultural representations and signifying practices*. SAGE Publications.

Hanh, Thich Nhat (1998). *The heart of the Buddha's teaching: Transforming suffering into peace, joy, and liberation.* Broadway Books.

Hard, T. (2014). *The role of ideology in the perception of information and its consequences for political behavior* . In S. L. M. Street & E. T. M. M. De Groote (Eds.), *The psychology of political behavior* (pp. 175-194). Routledge.

Harman, G. (1975). Moral Relativism Defended. *The Philosophical Review*.

Harman, G. (2011). The idea of a human right: A philosophical perspective. *Social Philosophy and Policy*, *28*(1), 1-29

Harman, G. (1975). Moral Relativism Defended. *Philosophical Review*, *84*(1), 3–22.

Harris, D. (2014). *Waking up: A guide to spirituality without religion*. Simon & Schuster.

Hawkins, J. (2021). *A thousand brains: A new theory of intelligence*. Basic Books.

Heffernan, Margaret. (2011). *Willful blindness: Why we ignore the obvious at our peril*. Walker Publishing Company.

Hofstede, G. (1980). *Culture's consequences: International differences in work-related values*. Sage Publications.

Hood, R. W., & Hill, P. C. (2020). *The psychology of religion: An empirical approach*. Guilford Press.

Hubel, D. H., & Wiesel, T. N. (1962). Receptive fields, binocular interaction and functional architecture in the cat's visual cortex. *Journal of Physiology, 160* (1), 106-154. https://doi.org/10.1113/jphysiol.1962.sp006837

Hume, D. (1757/2008). Of the standard of taste. In D. Wiggins & D. Clarke (Eds.), *Philosophy of art: Readings ancient and modern* (pp. 247-258). Wiley-Blackwell.
[Originally published 1757.]

Husserl, E. (1970). *The crisis of European sciences and transcendental phenomenology* (D. Carr, Trans.). Northwestern University Press. (Original work published 1936)

Ioannidis, J. P. A. (2005). Why most published research findings are false. *PLoS Medicine*, *2*(8), e124.

James, W. (1902). *The varieties of religious experience: A study in human nature*. Longmans, Green, and Co.

James, W. (1907). *Pragmatism: A new name for some old ways of thinking*. Longmans, Green & Co.

Janis, I. L. (1972). *Victims of groupthink: A psychological study of policy decisions and fiascoes*. Houghton Mifflin.

Janis, I. L. (1982). *Groupthink: Psychological studies of policy decisions and fiascoes*. Houghton Mifflin.

Johnson, D. D. P., & Fowler, J. H. (2011). The evolution of overconfidence. *Nature*, *477*(7364), 317-320. https://doi.org/10.1038/nature10380

Johnson-Laird, P. N. (1983). *Mental models: Toward a cognitive science of language, inference, and consciousness*. Harvard University Press.

Jost, J. T., & Hunyady, O. (2003). Fundamental ideological motives and the protection of the status quo: The system justification model. *Psychological Inquiry*, *14*(1), 89-103. https://doi.org/10.1207/S15327965PLI1401_8

Jung, C. G. (1959). Aion: Researches into the phenomenology of the self. Princeton University Press.

Justice, C., Rice, J., Warry, W., Inglis, S., & Miller, S. (2009). Academic skill development—Inquiry-based learning. *Canadian Journal of Higher Education, 39*(2), 129–155.

Kabat-Zinn, J. (2005). *Wherever you go, there you are: Mindfulness meditation in everyday life*. Hachette Books.

Kabat-Zinn, J. (2013). *Full catastrophe living: Using the wisdom of your body and mind to face stress, pain, and illness* (Revised and updated). Bantam.

Kahan, D. M., et al. (2017). The tragedy of the risk-perception gap in the regulation of environmental risks. *Risk Analysis*, *37*(1), 71-82. doi:10.1111/risa.12620.

Kahneman, D. (2011). *Thinking, fast and slow*. Farrar, Straus and Giroux.

Kandel, E. R., Schwartz, J. H., & Jessell, T. M. (2013). *Principles of Neural Science* (5th ed.). McGraw-Hill.

Kant, I. (1785). *Groundwork of the metaphysics of morals* (Trans. M. Gregor). Cambridge University Press

Kaplan, S. (1995). The restorative benefits of nature: Toward an integrative framework. *Journal of Environmental Psychology*, *15*(3), 169-182. https://doi.org/10.1016/0272-4944(95)90001-2

Katzenbach, J. R., & Smith, D. K. (1993). *The wisdom of teams: Creating the high-performance organization*. HarperBusiness.

Kegan, R. (1982). *The evolving self: Problem and process in human development.* Harvard University Press.

Kelly, G. A. (1955). *The psychology of personal constructs. Volume 1: A theory of personality*. Norton.

Kolb, D. A. (1984). *Experiential learning: Experience as the source of learning and development*. Prentice-Hall.

Kölbel, M. (2003). Faultless Disagreement. *Mind*, *112*(448), 675–703.

Kölbel, M. (2004). Faultless disagreement. *Proceedings of the Aristotelian Society*, *104*: 53–73.

Krizhevsky, A., Sutskever, I., & Hinton, G. E. (2012). Imagenet classification with deep convolutional neural networks. *Advances in Neural Information Processing Systems, 25*, 1097–1105. https://proceedings.neurips.cc/paper_files/paper/2012/file/c399862d3b9d6b76c8436e924a68c45b-Paper.pdf

Kuhn, T. S. (1962). *The structure of scientific revolutions*. University of Chicago Press.

Kunda, Z. (1990). *The case for motivated reasoning. Psychological Bulletin, 108*(3), 480-498. doi:10.1037/0033-2909.108.3.480

Kurzban, R., & Aktipis, C. A. (2007). Modularity and the evolution of social behavior. *Journal of Evolutionary Psychology*, 5(3-4), 114-130.

Lakatos, P., Karmos, G., Mehta, A. D., Ulbert, I., & Schroeder, C. E. (2008). Oscillatory entrainment of neuronal ensembles as a mechanism of attentional selection. *Science, 320* (5872), 110-113. https://doi.org/10.1126/science.1154735

Latané, B., & Darley, J. M. (1970). The unresponsive bystander: Why doesn't he help? *Psychology Today, 5*(1), 58-61

Latour, B. (2005). *Reassembling the social: An introduction to actor-network-theory*. Oxford University Press.

Latour, B., & Woolgar, S. (1979/1986). *Laboratory life: The construction of scientific facts* (2nd ed.). Princeton University Press.

LeDoux, J. E. (1996). *The emotional brain: The mysterious underpinnings of emotional life*.

Lee, J. H., & Whang, M. (2017). Self-deception and its mediated effects on romantic relationship satisfaction. *Journal of Social and Personal Relationships, 34*(7), 1053-1075.

Leman, P. J. (2013). The illusion of the concrete: The misconception of implicit bias. *Frontiers in Psychology, 4*, 1-6.

Leman, P. J., & Cinnirella, M. (2007). A major event has a major cause: Evidence for the role of heuristics in reasoning about conspiracy theories . *Social Psychological Review*, *9*(2), 18-28.

Levin, D. T., & Simons, D. J. (1997). Change blindness as a new interpretive framework for studying visual perception. *Proceedings of the National Academy of Sciences*, *94*(1), 134-148.

Lilienfeld, S. O., Ammirati, R., & Landfield, K. (2009). Giving Debiasing Away: The Effects of a Public Information Campaign on the Reduction of Cognitive Bias. *Psychological Science*, *20*(3), 198-200. https://doi.org/10.1111/j.1467-9280.2009.02285.x

Locke, E. A., & Latham, G. P. (1990). Goal setting: Theory and practice. *Organizational Behavior and Human Decision Processes*, *50*(2), 212–247. https://doi.org/10.1016/0749-5978(90)90018-I

Locke, E. A., & Latham, G. P. (2002). Building a practically useful theory of goal setting and task motivation: A 35-year odyssey. *American Psychologist*, *57*(9), 705-717. https://doi.org/10.1037/0003-066X.57.9.705

Loewenstein, G., O'Donoghue, T., & Rabin, M. (2003). Projection bias in predicting future utility. *The Quarterly Journal of Economics*, *118*(4), 1209–1248.

Lord, C. G., Ross, L., & Lepper, M. R. (1979). Biased assimilation and attitude polarization: The effects of prior theories on subsequently considered evidence. *Journal of Personality and Social Psychology*, *37*(11), 2098-2109.

Lorenz, K. (1981). The companion of the animal. In *Studies in animal and human behavior* (Vol. 2). Academic Press.

Lukes, S. (2005). Power: A radical view. (2nd ed.)Palgrave Macmillan.

Lutz, A., Slagter, H. A., Dunne, J. D., & Davidson, R. J. (2008). Attention regulation and monitoring in meditation. *Trends in Cognitie Sciences, 12*(4), 163-169. https://doi.org/10.1016/j.tics.2008.01.005

MacFarlane, J. (2014). *Assessment sensitivity: Relative truth and its applications*. Oxford University Press.

Mack, A., & Rock, I. (1998). *Inattentional blindness*. MIT Press

Marslen-Wilson, W. D. (1987). Functional parallelism in spoken word recognition. *Cognitive Psychology* , *19*(3), 273-292. https://doi.org/10.1016/0010-0285(87)90023-5

Maslow, a. (1964). Religions, values, and peak-experiences.

McCombs, M., & Shaw, D. L. (1972). The agenda-setting function of mass media. *Public Opinion Quarterly, 36*(2), 176–187.

McGilchrist, I. (2009). The master and his emissary: The divided brain and the making of the Western world. Yale University Press.

McKay, R., & Dennett, D. C. (2009). The evolution of belief in the supernatural. *Proceedings of the Royal Society Biological Sciences*, 276(1667), 127-133. https://doi.org/10.1098/rspb.2008.0847

Mead, Margaret. (1928). Coming of age in Samoa. William Morrow & Company.

Merleau-Ponty, M. (1962). *Phenomenology of perception* (C. Smith, Trans.). Routledge. (Original work published 1945)

Merry, S. E. (1988). Legal pluralism. *Law & Society Review*, *22*(5), 869–896.

Mezirow, J. (1991). *Transformative dimensions of adult learning.* Jossey-Bass.

Millar, M. G., & Millar, K. (2021). The Ethics of Omission: A Study of Omission Bias in Practical Decision-Making. *Journal of Organizational Behavior*, *42*(8), 1276-1298.

Miller, D. T., & Ross, M. (1975). Self-serving biases in the attribution of causality: Fact or fiction? *Psychological Bulletin, 82*(2), 213–225.

Milner, A. D., & Goodale, M. A. (2006). *The visual brain in action* (2nd ed.). Oxford University Press.

Moray, N. (1969). *Attention: Selective processes in vision and hearing*. Academic Press.

Nagel, T. (1974). What is it like to be a bat? *The Philosophical Review* , *83*(4), 435-450. https://doi.org/10.2307/2183914

Neff, K. (2011). *Self-compassion: The proven power of being kind to yourself.* William Morrow.

Neisser, U. (1979). *Cognitive Psychology*. Appleton-Century-Crofts. ISBN: 978-0818501093.

Nesse, R. M. (1990). Evolutionary explanation of emotions. *Human Nature, 1*(3), 261-289. https://doi.org/10.1007/BF02734103

Nickerson, R. S. (1998). Confirmation bias: A ubiquitous phenomenon in many guises. *Review of General Psychology, 2*(2), 175-220.

Niemi, R. G., & Jennings, M. K. (1991). Issues and inheritance in the formation of party identification. *American Journal of Political Science, 35*(4), 970–988.

Nietzsche, F. (1989). *On the genealogy of morals*. (W. Kaufmann & R. J. Hollingdale, Trans.). Vintage Books. (Original work published 1887).

Nisbett, R. E., & Masuda, T. (2003). Culture and perception. *The American Psychologist, 58*(5), 269-282. https://doi.org/10.1037/0003-066X.58.5.269

Noë, A. (2004). *Action in perception*. MIT Press.

Noelle-Neumann, E. (1974). The spiral of silence: A theory of public opinion. *Journal of Communication, 24*(2), 43-51. https://doi.org/10.1111/j.1460-2466.1974.tb00367.x

Norgaard, K. M. (2011). Climate denial: Emotion, reason, and collective responsibility in a changing climate. *Cultural Studies, 25*(2), 204-223. https://doi.org/10.1080/09502386.2011.555627

Norman, D. A. (2013). *The design of everyday things* (Revised and expanded edition). Basic Books.

Nussbaum, M. C. (1997). *Cultivating humanity: A classical defense of reform in liberal education*. Harvard University Press.

Ochsner, K. N., & Gross, J. J. (2005). The cognitive control of emotion. *Trends in Cognitive Sciences, 9*(5), 242–249.

Oettingen, G., & Gollwitzer, P. M. (2010). Strategies for goal setting and goal striving: Motivational processes for a better life. *In Handbook of Health Psychology* (pp. 1-25). Academic Press. https://doi.org/10.1016/B978-0-12-374283-6.00001-2

Omer, H., & Alon, N. (1994). The continuity principle: A unified approach to disaster and trauma. *American Journal of Community Psychology, 22*(2), 273-287.

Ovadia, Y., Raghavan, M., Head, T., Algazi, A., & Vranceanu, A. (2019). Can you trust your model's uncertainty? A case study of distributional shift." *arXiv preprint arXiv:1908.07577*. Retrieved from https://arxiv.org/abs/1908.07577

Palmer, P. J. (1998). *The courage to teach: Exploring the inner landscape of a teacher's life*. Jossey-Bass.

Parasuraman, R., & Hsu, J. (2000). A neuroergonomic approach to the design of work with the human brain in mind. *Technology and Human Factors, 1*(1), 1-19. https://doi.org/10.1007/s10479-005-6068-2

Pariser, E. (2011). *The filter bubble: What the internet is hiding from You* Penguin.

Paul, R., & Elder, L. (2014). *Critical thinking: Tools for taking charge of your professional and personal life* (3rd ed.). Pearson.

Peirce, C. S. (1877). The fixation of belief. *Popular Science Monthly*, *12*, 1-15

Peirce, C. S. (1903). Pragmatism and Pragmaticism. In C. Hartshorne & P. Weiss (Eds.), *Collected Papers of Charles Sanders Peirce*, vol. 5. Harvard University Press

Pennebaker, J. W. (1997). *Opening up: The healing power of expressing emotions*. Guilford Press.

Piaget, J. (1971). *Psychology and epistemology*. Penguin Books.

Piaget, J. (1978). *Structuralism*. Harper Torchbooks.

Plato, "Apology" 36c-38a

Plato. (2003). *Theaetetus*. (M.J. Levett, Trans.). In J. Cooper (Ed.), *Plato: Complete works* (pp. 157–234). Hackett Publishing Company. (Original work published c. 369 B.C.E.)

Plous, S. (1993). *The psychology of judgment and decision making*. McGraw-Hill.

Posner, M. I. (1980). Orienting of attention. *Quarterly Journal of Experimental Psychology, 32*(1), 3-25. https://doi.org/10.1080/00335558008248231

Posner, M. I., & Petersen, S. E. (1990). The attention system of the human brain. *Annual Review of Neuroscience*, *13*(1), 25-42. https://doi.org/10.1146/annurev.ne.13.030190.000325

Posner, M. I., & Rothbart, M. K. (2007). Attention and self-regulation: Control, adaptability, and performance. *Cognitive, Affective, & Behavioral Neuroscience*, *7*(2), 137-143. https://doi.org/10.3758/CABN.7.2.137

Pound, R. (1912). The scope and purpose of sociological jurisprudence (Part III). *Harvard Law Review*, *25*(8), 489–516.

Quine, W. V. O. (1951). Two dogmas of empiricism. *The Philosophical Review*, *60*(1), 20-43.

Radford, A., Wu, J., Child, R., Luan, D., Amodei, D., & Sutskever, I. (2019). *Language Models are Unsupervised Multitask Learners*. OpenAI. Retrieved from https://cdn.openai.com/transcripts/LanguageModelsAreUnsupervisedMultitaskLearners.pdf

Raine, A., & Chen, Q. (2017). Neurobiological factors in the development of antisocial behavior: A review of literature. *Journal of Abnormal Psychology*, 126(7), 783-796. https://doi.org/10.1037/abn0000334

Rappaport, Roy A. (1999). Ritual and religion in the making of humanity. Cambridge University Press.

Rayner, K., & Castelhano, M. S. (2008). Eye movements during reading and visual search: The role of attention and memory. In *Visual Attention* (pp. 243-263). Psychology Press. https://doi.org/10.4324/9780203809685

Rensink, R.A., O'Regan, J.K., & Clark, J.J. (1997). To see or not to see: The need for attention to perceive changes in scenes. *Psychological Science*, *8*(5), 368–373.

Rorty, R. (1979). *Philosophy and the mirror of nature*. Princeton University Press.

Ross, M., & Sicoly, F. (1979). Egocentric biases in availability and attribution. *Journal of Personality and Social Psychology* , *37*(3), 322–336.

Ross, M., & Sicoly, F. (1979). Egocentric biases in availability and attribution. *Journal of Personality and Social Psychology* , 37(3), 322–336.

Rumsfield, D. (February 12, 2002). U. S. Department of Defense News Briefing.

Sackeim, H. A., & Gur, R. C. (1978). The effects of self-deception on the perception of body image. *Journal of Abnormal Psychology*, *87*(5), 589-595. https://doi.org/10.1037/0021-843X.87.5.589

Salvucci, D. D., & Taatgen, N. A. (2008). The role of attention in multitasking: A cognitive architecture for the dual-task paradigm. *Psychological Review*, 115(3), 778-792. https://doi.org/10.1037/0033-295X.115.3.778

Samuelson, W. & Zeckhauser, R. (1988). Status quo bias in decision making. *Journal of Risk and Uncertainty*, *1*(1), 7-59.

Sapir, E. (1929). The status of linguistics as a science. *Language*, *5*(4), 207–214.

Sartre, J.-P. (2003). *Being and nothingness* (Barnes, Trans.). Routledge. (Original work published 1943)

Schwartz, G. E., & Davidson, R. J. (2002). The role of consciousness in emotional experience. In R. J. Davidson, K. R. Scherer, & H. H. Goldsmith (Eds.), *Handbook of affective sciences* (pp. 681-700). Oxford University Press.

Searle, J. R. (1983). *Intentionality: An essay in the philosophy of mind.* Cambridge University Press.

Seligman, M. E. P., & Csikszentmihalyi, M. (2000). Positive psychology: An introduction. *American Psychologist*, *55*(1), 5-14.

Sharot, T. (2011). The optimism bias. *Current Biology, 21* (23)

Siegel, D. J. (2010). *Mindsight: The new science of personal transformation.* Bantam.

Simons, D. J., & Chabris, C. F. (1999). Gorillas in our midst: Sustained inattentional blindness for dynamic events. *Perception*, *28*(9), 1059-1074. https://doi.org/10.1068/p281059

Simons, D. J., & Levin, D. T. (1997). Change blindness. *Trends in Cognitive Sciences*, *1*(7), 261–267.

Simons, D. J., & Rensink, R. A. (2005). Change blindness: Perception and attention. In *The Oxford Handbook of Cognitive Psychology* (pp. 155-173). Oxford University Press

Sinek, S. (2009). *Start with why: How great leaders inspire everyone to take action*. Penguin Group.

Singer, J. A., & Salovey, P. (1993). The rememberer and the storyteller: The role of autobiographical memory in self-definition. In *The Self: Interdisciplinary Approaches* (pp. 14-43). Wiley.

Singer, P. (1993). *Practical Ethics* (2nd ed.). Cambridge University Press.

Steg, L., & Vlek, C. (2009). *Encouraging sustainable behavior: An overview of concepts and evidence. Journal of Social Issues*, *65*(1), 1-11.

Stroud, N. J. (2011). *Niche news: The politics of news choice*. Oxford University Press.

Sunstein, C. R. (2009). *Going to extremes: How like minds unite and divide*. Oxford University Press.

Sweller, J. (1988). Cognitive load during problem solving: Effects on learning. *Cognitive Science* , *12*(2), 257-285. https://doi.org/10.1207/s15516709cog1202_4

Tajfel, H. & Turner, J. C. (1979). An integrative theory of intergroup conflict. In W. G. Austin and S. Worchel (Eds.), *The social psychology of intergroup relations*, pp. 33–47. Brooks/Cole.

Tajfel, H. (1982). *Social identity and intergroup relations*. Cambridge University Press.

Tajfel, H. (1982). Social identity and intergroup relations. In H. Tajfel (Ed.), *Social Identity and Intergroup Relations* (pp. 1-39). Cambridge University Press.

Tamanaha, B. Z. (2000). A non-essentialist version of legal pluralism. *Journal of Law and Society*, *27*(2), 296–321.

Taylor, C. (1985). The explanation of action. In *Philosophy and the human sciences: Philosophical papers 2*. Cambridge University Press.

Taylor, C. (Ed.) (1985). *Philosophy and the human sciences: Philosophical papers 2*. Cambridge University Press.

Thaler, R. H. (1991). *Quasi-rational economics*. Russell Sage Foundation.

Thaler, R. H., & Larrick, R. P. (2000). The effects of the framing of time on decisions. *In A. Tversky & D. Kahneman (Eds.), Advances in prospect theory: Cumulative potential and decision making* (pp. 355-376). Cambridge University Press.

Thaler, R. H., & Sunstein, C. R. (2008). *Nudge: Improving decisions about health, wealth, and happiness*. Yale University Press.

The Dalai Lama (1997). *The four noble truths*. Thorsons.

Tilly, C. (1999). *Durable inequality*. University of California Press.

Tolle, E. (1999). *The power of now: A guide to spiritual enlightenment*. Namaste Publishing.

Treisman, A. (1986). Properties and objects in visual perception. *Journal of Experimental Psychology: Human Perception and Performance*, *12*(1), 3-16. https://doi.org/10.1037/0096-1523.12.1.3

Treisman, A. M. (2006). Attention and the integration of features. *Perception & Psychophysics, 63*(2), 256-263. https://doi.org/10.3758/BF03193515

Trivers, R. (2011). *The folly of fools: The logic of deceit and self-deception in human life.* Basic Books.

Turner, J. C., & Hogg, M. A. (1987). *Rediscovering the Social Group: A Self-Categorization Theory*. Blackwell Publishing.

Turner, R., & Bennet, A. (2025). *Knowledge capacities II: Cultivating infinite learning*. MQIPress.

Tversky, A. & Kahneman, D. (1974). Judgment under uncertainty: Heuristics and biases. *Science, 185*(4157), 1124-1131.

Tversky, A., & Kahneman, D. (1973). Availability: A heuristic for judging frequency and probability. *Cognitive Psychology*, *5*(2), 207-232.

van der Kolk, B. A. (2014). *The Body Keeps the Score: Brain, Mind, and Body in the Healing of Trauma*. Viking. ISBN: 978-0143127741.

Vaswani, A., Shazeer, N., Parmar, N., Uszkoreit, J., Jones, L., Gomez, A. N., ... & Polosukhin, I. (2017). Attention is all you need. In *Advances in neural information processing systems* (pp. 5998-6008).

Vaughan, D. (1996). *The challenger launch decision: Risky technology, culture, and deviance at NASA*. University of Chicago Press.

Vazire, S., & Carlson, E. N. (2011). Self-knowledge of personality: Do people know themselves as well as others know them? *Journal of Personality and Social Psychology*, *101*(2), 202-216. https://doi.org/10.1037/a0021754

von Hippel, W., & Trivers, R. L. (2011). The evolution and functions of self-deception. *Behavioral and Brain Sciences*, *34*(2), 103-104. https://doi.org/10.1017/S0140525X1000160X

von Neumann, John, and Oskar Morgenstern (1944). Theory of games and economic behavior. Princeton University Press

Vygotsky, L. S. (1986). Thought and language. (Alex Kozulin, Ed.). MIT Press.

Weinstein, N. D. (1980). Unrealistic optimism about future life events. *Journal of Personality and Social Psychology*, *39*(5), 806-820.

Whorf, B. L. (1956). Language, thought, and reality: Selected writings of Benjamin Lee Whorf (John B. Carroll, Ed.). MIT Press.

Wickens, C. D., Lee, J., Liu, Y., & Becker, S. (2015). *An introduction to human factors engineering* (2nd ed.). Pearson.

Williams, B. (1973). Deciding to believe. In *Problems of the Self* (pp. 136–151). Cambridge University Press.

Williams, B. (Ed.) (1973). *Problems of the self: Philosophical papers.* Cambridge University Press.

Wilson, E. O. (1998). *Consilience: The unity of knowledge*. Alfred A. Knopf.

Wilson, T. D., & Dunn, E. W. (2004). Self-knowledge: Its limits, and ways to overcome them. *Social Justice Research*, 17(2), 253-270. https://doi.org/10.1007/s11211-004-5054-2

Winch, P. (1958). *The idea of a social science and its relation to philosophy*. Routledge & Kegan Paul.

Wong, D. B. (1984). *Moral relativity*. University of California Press.

Yuille, A. L., & Kerkhoff, A. (2006). Statistical models for change detection and related human perception. *Proceedings of SPIE - The International Society for Optical Engineering*, 6063, 60630H.

Yuille, A. L., & Kersten, D. (2006). Vision as Bayesian inference: Analysis by synthesis? *Trends in Cognitive Sciences, 10*(7), 301-308. https://doi.org/10.1016/j.tics.2006.05.002

Zaller, J. R. (1992). *The nature and origins of mass opinion*. Cambridge University Press.

Zeiler, M. D., & Fergus, R. (2014). Visualizing and understanding convolutional networks. In *European Conference on Computer Vision (ECCV),* 818–833. https://arxiv.org/abs/1311.2901

Zerubavel, E. (1995). *Social Mindscapes: An invitation to cognitive sociology*. Harvard University Press.

Zimmerman, B. J. (2002). Becoming a self-regulated learner: An overview. *Theory Into Practice, 41*(2), 64-70. https://doi.org/10.1207/s15430421tip4102_2

About the Author

Alex Bennet is a Professor, Innovation and Knowledge Institute Southeast Asia (IKI-SEA), Bangkok University, and the Director of the Mountain Quest Institute, a research and retreat center located in the Allegheny Mountains of West Virginia. Through three quests—the quests for knowledge, consciousness and meaning—the Institute is dedicated to helping individuals achieve personal and professional growth, and organizations create and sustain high performance in a rapidly changing, uncertain, and increasingly complex world. Alex is the former Chief Knowledge Officer and Deputy CIO for Enterprise Integration of the U.S. Department of the Navy, having previously served as Acquisition Reform Executive and Standards Improvement Executive, and is recipient of the Distinguished Public Service Award, the highest civilian honor from the Secretary of the Navy. She has published hundreds of papers and journal articles, and authored, co-authored or edited over 40 books, primarily with her life partner, Dr. David Bennet, a nuclear physicist and neuroscientist, who recently transitioned. Together, the Drs. Bennet have spoken and taught around the world. This is the 3rd book in the series looking at the current state of the world through different lenses. These are *fRAGmenttion: The Power of e Pluribus Unum* and *The Taint of Hubris: Rediscovering Humility and Bridging Authentic Connection*. A fourth book: *Becoming **W**ise, **O**pen, **K**ind, and **E**mpowered* is an open source volume downloadable from ResearchGate and www.mqresearchcenter.com Other 2025 publications include *Infinite Learning: Your Life, Your Choice*; *C&C Thinking: Becoming Whole* (creative and critical thinking); and *Contiguity: Entangled Living and Learning*. Alex believes in the multidimensionality and interconnectedness of humanity as we move out of infancy into full consciousness. Contact her at alex@mountainquestinstitute.com

The Mountain Quest Research Center, located in the Allegheny Mountains of West Virginia, is a research, retreat, and learning center dedicated to helping individuals achieve personal and professional growth, and organizations create and sustain high performance in a rapidly changing, uncertain, and increasingly complex world. MQI has three quests: the Quest for Knowledge, the Quest for Consciousness, and the Quest for Meaning. MQI is scientific, humanistic, and spiritual and finds no contradiction in this blend.

Eye Opening Research for Today's World!

fRAGmented: e Pluribus Unum (2025)

Alex Bennet with Foreword by YOU

This is our world, a tapestry woven with threads of diversity and division. As we journey in this book through the complexities of our world, we critically explore fragmentation in the physical, holistic human, digital, narrative/art, and societal domains, and delve into the societal political fragmentation occurring today.

The Taint of Hubris: (2025)

Rediscovering humility and bridging authentic connections

Alex Bennet with Foreword by William Halel

Characterized by excessive pride or arrogance—and referring to something being spoiled, sullied, or negatively influenced in some way—hubris can be thought of as a contaminating aspect of human nature. It stands out as a pervasive taint that has woven itself into mythology, literature and history, and is highly visible in the societal political landscape of today.

Becoming Wise, Open, Kind, Empowered (2025)

The Millenium Challenge

Alex Bennet

For many, being "woke" signifies a positive awareness and active engagement with social justice issues, which is the intent of social and cultural awakening consistent with spiritual awakening However, in this fragmented world, core tenets of democracy such as *diversity, inclusiveness and equality* are being tossed aside and translated through prejudicial and biased beliefs and political frames. If we are wise enough to think in Whole Thought and consult and understand the patterns reflecting human decisions and events, this has been historically proven questionable. And herein lies the purpose of this book.

Whole Thought:
The Rise of Human Intelligence (2024)

Alex Bennet and Robert Turner with Foreword by David Bennet

In tracing the arc of human progress, one can discern an underlying pattern steering the course of our intellectual evolution—a gradual but persistent gravitation towards what is now identified as Whole Thought. This paradigm does not represent a mere milestone, but rather signifies a shift, a transformation in the fabric of cognition brought into relief by the cumulative ascent of human intelligence.

The journey through the landscape of Whole Thought reveals an integrative approach to cognition that is as timely as it is timeless. It beckons us to rethink how we perceive, learn, and engage with the world and each other. Whole Thought is more than a theoretical concept; it is a call to action—a framework for living, learning, and leading in a manner that is reflective, inclusive, and deeply interconnected.

TRIBUTE

Deep appreciation for the life of

David Hughes Bennet

1934 – 2025

Author, Friend, Infinite Learner

Our capacity for knowledge is not a fixed quantity but a horizon ever-expanding. It is the reservoir from which we draw strength, the lens through which we view possibility, and the compass by which we navigate the future. It is the creation of the rise of human intelligence.

Knowledge Capacities: Igniting Whole Thought—offering 40 Knowledge Capacities in support of Whole Thought, is open source and downloadable as a PDF from www.ResearchGate.edu , www.mountainquestinn.com (scroll to the bottom of the first page) and www.mqresearchcenter.com In cooperation with ROM Knowledge Management and for ease of copying and sharing, the Knowledge Capacities in the entire Whole Thought set are individually provided at www.kmrom.com/knowledge-capacities for download. By way of extension, Whole Thought and Knowledge Capacities are introductions to the new body of Human and AI Thoughtware.

Knowledge Capacities II: Cultivating Infinite Learning—offering 24 additional Knowledge Capacities in support of Whole Thought and cultivating infinite learning.

In today's dynamic and rapidly evolving environment, fostering capacity has become increasingly essential. Capacity refers to the broad potential or inherent ability of individuals and organizations to learn, adapt, and grow over time. It encompasses the fundamental ways of thinking, being, and acting that allow one to effectively engage with dynamic and complex environments. The broad and adaptive nature of capacities provides a robust foundation for continuous evolution and success.

Infinite Learning: Your Life, Your Choice (2025)
by Alex Bennet with Foreword by Vincent Ribiére

Infinite learning is the pulse of human existence, the essence that breathes life into our quest for understanding, innovation, and growth. Embracing infinite learning in pursuit of Whole Thought ensures that we do not merely exist but truly live, constantly expanding our horizons and discovering new potentials. *Living is learning; learning is living.*

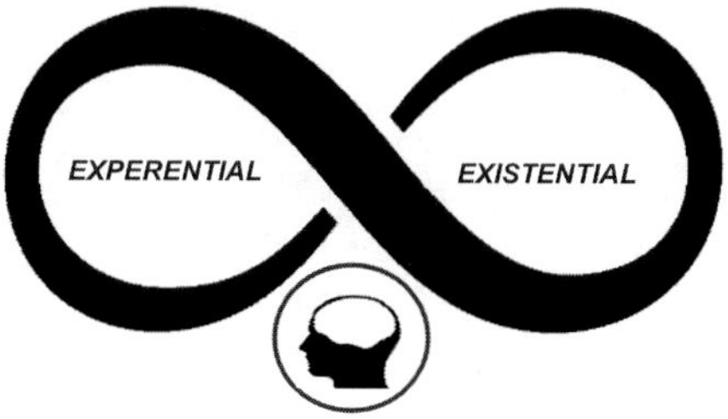

Human and AI Thoughtware

Unleashing the Human Mind
A Consilience Approach to Managing Self

Whole Thought
The Rise of Human Intelligence

Infinite Learning
Your Life, Your Choice

Knowledge Capacities I
Igniting Whole Thought

Knowledge Capacities II
Cultivating Infinite Learning

Reblooming the Knowledge Movement
The Democratization of Organizations`

C&C Thinking: Becoming Whole (2025) Critical and Creative

Alex Bennet and Robert Turner
with Foreword by Moria Levy

In an era marked by rapid technological advancement and constant change, the ability to think critically and creatively is more crucial than ever. As we look toward the future, it becomes evident that the traditional reliance on past patterns to predict and plan for what lies ahead is insufficient.

Contiguity: Entangled Living and Learning (2025)

Alex Bennet with Foreword by Chulatep Senivongse

Learning and living are contiguous experiences, with mind creating the subjective relationships that create the temporal and spatial relationships in our stories and memories. Our thoughts, sensations, and perspectives form the connected and cohesive experience of the contiguous mind. Embrace the entangled dance of living and learning, and discover the profound connections that define our shared existence.

Innovative Creativity: (2024) Creating with Innovation in Mind

Alex Bennet and Arthur Shelley with Charles Dhewa
with Foreword by Robert Turner

More than ever, how do we release the Genie from the lamp? How do we tap the next level of creativity and innovation that we need here on Planet Earth? This groundbreaking work beckons us to deepen our innate creativity capacities in a new and expansive way to summon the genius within each of us.

Recent Offerings from MQIPress

Reblooming the Knowledge Movement: The Democratization of Organizations (2023)

Alex Bennet and Robert Turner with Foreword by Rory Chase and chapters contributed by Francisco Javier Carrillo, Mark Boyes, Florin Gaiseanu, Chulatep Senivongse, and Milton de Sousa

Reblooming unfolds on millennia of human challenges and advances. Now, at every level and at every reach across organizations, networks, and nations there is a new coalescing of democratization, intelligent learning, and capacity for surmounting complexity. With freedom of thought, freedom of expression, and freedom of association, ideas beget ideas. The emergent result is the rich globalization of knowledge and its close companion, innovation.

INside INnovation: Looking from the Inside Out (2023)

Dr. Alex Bennet and Dr. Rajat Bais (Eds.) with Prologue by Dr. Leif Edvinsson and Foreword by Dr. Massimo Pregnolato

In addition to a collection of insightful innovation case studies, this book offers an unusual look at creativity and innovation from the inside out. Three innovators – a scientist, an organizational guru, and an artist – share the personal passions that have driven their success. "And, then, looking from the inside out, readers are provided the opportunity to evaluate their own organizations against the Most Innovative Knowledge Organization (MIKE) international study program and awards criteria, thus engaging their own innovative juices."

Unleashing the Human Mind: A Consilience Approach to Managing Self (2022)

David Bennet, Alex Bennet, Robert Turner
with Foreword by Florin Gaiseanu

What does it mean to be human? Increasingly, we recognize that we are infinitely complex beings with immense emotional and spiritual, physical and mental capacities. Presiding over these human systems, our brain is a fully integrated, biological, and extraordinary organ that is preeminent in the known Universe. Its time has come. This book is grounded in the Intelligent Complex Adaptive Learning System (ICALS) theory based on over a decade of researching experiential learning through the expanding lens of neuroscience.

Review Bites from Around the World:

Once in a while, I am exposed to a work so profound that it literally causes a massive shift in my own thinking and beliefs ... I found myself riveted to each paragraph as I embarked on a journey that vastly deepened my understanding of the learning process.

-Duane Nickull, Author, Technologist, and Seeker of Higher Truth, Canada

Every now and then a book comes along that compels your spirit. In these times of uncertainty and even great danger for humanity, this book reminds us of what it means to be human, our infinite potential and innate ability to learn and to love.

-Dr. Milton deSousa, Associate Professor, Nova School of Business and Economics, Portugal

Very few people have the gift to integrate such complex ideas, especially those about learning ... this work can be likened to the Webb Telescope, which gives us more clarity into our mysteries.

-Michael Stankosky, DSc, Author, Philosopher, Professor, Editor-Emeritus, Member of the Academy of Scholars, USA

And the *Unleashing Field Guide: An OrgZoo Quest*

Alex Bennet, Robert Turner, Arthur Shelley, Jane Turner and Mark Boyes (Illustrator) (2022)

Arthur Shelley's beloved OrgZoo critters – the voices in our heads – join us in a learning quest up the mountain to unleash the human mind.

Additional titles from MQIPress:

The Profundity and Bifurcation of Change: Parts I – V (2017; 2020)

Possibilities that are YOU! (a 22-volume conscious look book series) (2018)

The QUEST: Where the Mountains Meet the Library (2021)

Remembrance: Pathways to Expanded Learning (2020)

The Course of Knowledge: A 21st Century Theory (2015)

Decision-Making in The New Reality: Complexity, Knowledge and Knowing (2013)

Expanding the Self: The Intelligent Complex Adaptive Learning System (2015)

Leading with the Future in Mind: Knowledge and Emergent Leadership (2015)

With Passion, We Live and Love: Research, Prose, Verse and Music (2021)

Playing in the Mind Field: Volume 1: Life in the Field (2023)

Made in United States
North Haven, CT
04 November 2026

21766942R00163